With gathering murk—with muttering thunder and lambent shoots, we all duly awake,
South, north, east, west, inland and seaboard, we will surely awake.

—Walt Whitman, *Leaves of Grass*

Running the Numbers

an American self-portrait

To Brian,
with warm regards
from one shutterbug
to another.

chris jordan

2011

Introduction by Chris Bruce
Essays by Lucy R. Lippard and Paul Hawken

Museum of Art / Washington State University, Pullman
Prestel, Munich / Berlin / London / New York

CONTENTS

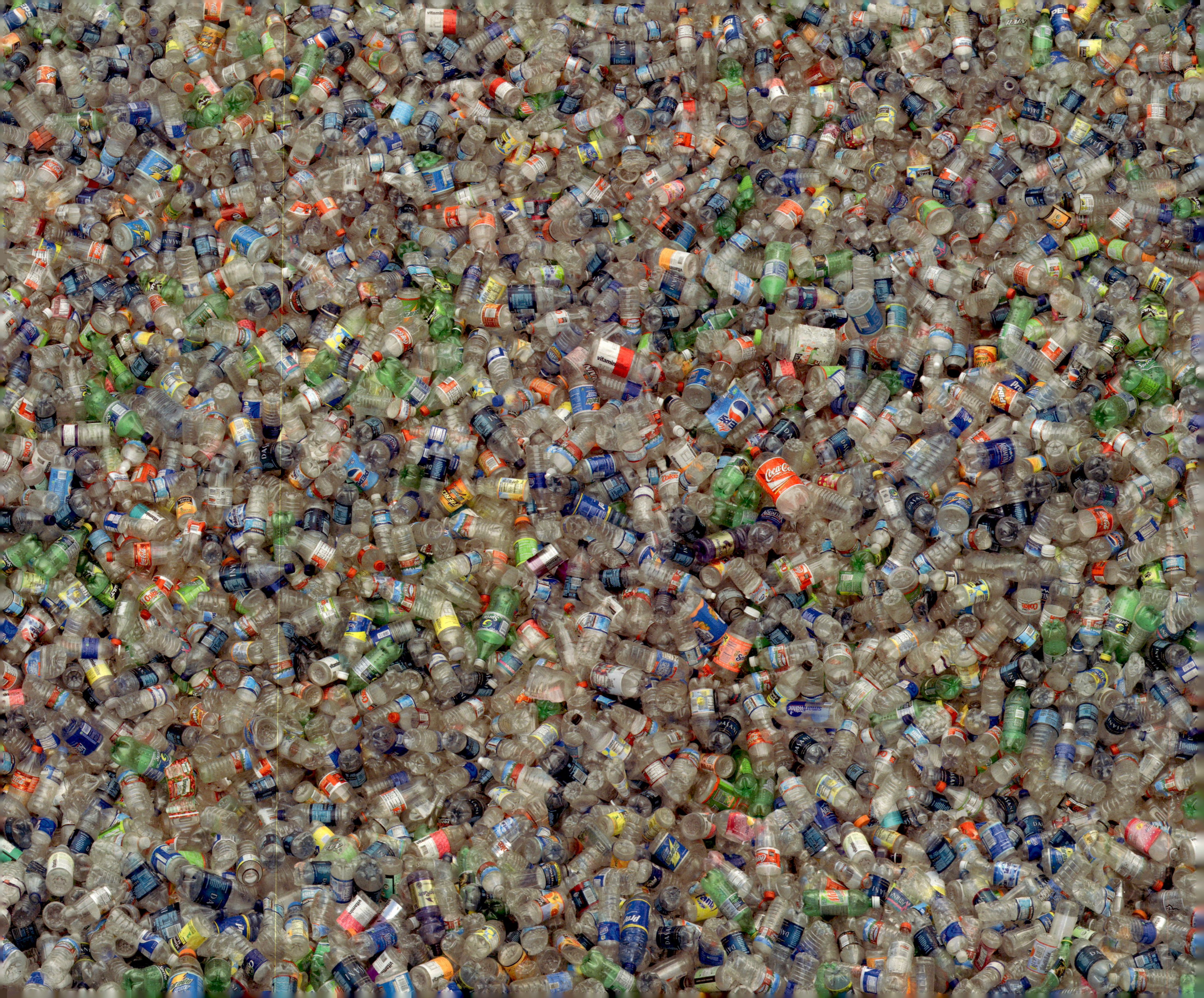
vitamin
Coca-Cola

Chris Bruce

CHRIS JORDAN: THE EFFECTIVE IMAGE

What can't be visualized can't be named.

Paul Hawken, *Blessed Unrest: How the Largest Movement in the World Came into Being and Why No One Saw It Coming* (2007)

From a distance, Chris Jordan's digitally fabricated landscape *Plastic Bottles* looks like a vast expanse of arctic tundra, yet from up close, we discover that it is a perfectly focused composition of two million plastic beverage bottles—the number used in the United States every five minutes (plate 6; fig. 1). You know how it goes. One day you see the photograph *Plastic Bottles*, and pretty soon, you see plastic bottles everywhere. Of course, they *are* everywhere. That's the point.

You begin to notice things: your neighbor's overflowing recycling bin; how often you are offered a bottle of water at a meeting; a spate of articles on how some specialty brands use the same tap water people in the Bronx or Los Angeles drink from their faucets. You just happen to see Lewis Black's comedy sketch on bottled water (which preposterously notes the number of transfats per bottle). A Sunday *Doonesbury* comic features "bottled air" ("Ahh, like a little whiff of Switzerland!" "Actually, we collect it in Newark, but who cares?").[1] You hear that the mayor of Los Angeles eliminated bottled water in city offices, and that the mayor of San Francisco is pushing tap water. (It turns out that tap water is actually safer than bottled water: it is tested for bacteria twenty-five times more often than bottled water.) Finally, you stop buying water in plastic bottles and you think, "Maybe plastic beverage bottles will be a nonissue by the time this exhibition is on view." Which raises a central question about Jordan's project: To what extent is his work dependent on the topicality of the subject matter?

Though topicality plays a role, the photographs also stand as autonomous works of art. The images are impressive. They are large—larger in scale than what most people expect when they think of a photograph. The work is attractive, and the compositions have a ring of familiarity, variously recalling a bamboo forest, a wheat field under a threatening Montana sky, a painting by Georges Seurat, a photograph by Ansel Adams. The details are clear and title information is concise and, in a way, horrifying. The overall effect is "efficient."

The sense of attraction and repugnance is so balanced that the two reactions are difficult to rectify in short order, except as transferred to the artist's technical prowess in creating these uncomfortable fictions. How he conjures these images becomes an initial point of focus. Jordan photographs individual objects or groupings of objects and then digitally stitches together thousands of the separate images into one massive accumulation. For *Plastic Bottles*, he photographed a cluster of a few hundred bottles in his driveway, then rearranged the configuration for another shot, and then another, each one taken on a slight angle

fig. 1: Plastic Bottles, 2007 (detail of plate 6)

to account for the vanishing-point perspective he wanted for the final picture. It took three weeks of such arrangements in the changing Seattle light to get a series of hundreds of photographs with consistent tonalities. He then wove them all together into an image of two million bottles in landscape-like perspective—an aesthetic strategy that he uses regularly. For *Cans Seurat* (plate 10; fig. 2), he photographed forty different aluminum cans from several different angles. Each of those photographs became a separate digital image file. Then, working from a high-resolution image of Seurat's famous painting *A Sunday on La Grande Jatte* (1884–86), Jordan used a computer program to pick out and place the cans like so many pointillist dabs of paint to create a huge photographic evocation of the Seurat painting composed of 106,000 cans, the equivalent of thirty seconds of our can consumption.

For the viewer, the technical elegance of Jordan's images initially serves to mask the overwhelming statistical references, and yet that same precision ultimately produces a visceral realization of the relationship between the one and the many. When the artist showed his work on the mock TV news show *The Colbert Report* in October 2007, there was a spontaneous gasp from the audience as a slide of the overall view of *Cans Seurat* shifted to a close-up. I think that gasp came from a palpable sense of seeing oneself in the big picture: We go from observing the overall image to identifying "my Pepsi" or "my V8."

fig. 2: Cans Seurat, 2007 (detail of plate 10)

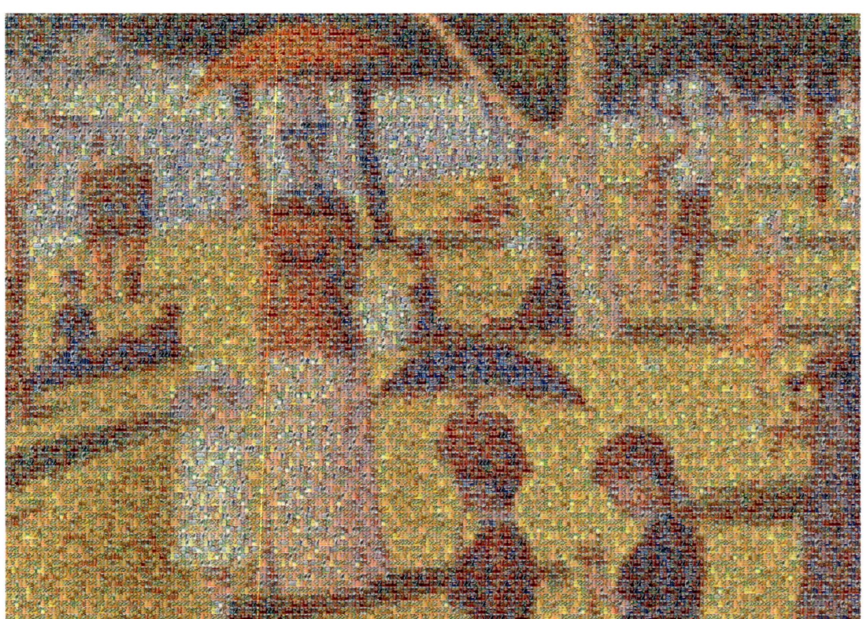

ART AND POLITICS

Given that art is never really immune to the social currents that move around it, are these works any different from previous art with a social conscience? The 1980s and 1990s saw an especially vigorous debate over the merits of political art. During this time, many politically inclined artists took on concerns theretofore championed by minority groups and by the American academe: racism, sexism, homophobia, Western hegemony, and the like. Many of these artists came out of art school during a period of increasing political sensitivity to identity and environmental issues. The Whitney Biennial of 1993 became the harbinger of this direction in art.

As well-meaning as it was, much of the work suffered from inflated expectations about the power of a social/political message over art qua art and, most unfortunately, from a simplistic division of the world into oppressors and victims. The too-easy political correctness of so much of the work led to widespread criticism, and the ultimate dilemma of these socially responsive works was that they never really left the art world or its institutions, even as they lobbied against elitism and privilege. Critic Robert Hughes wrote, "Such work is one long exercise in preaching to the choir. It affords the pleasure of striking more or less radical attitudes without the risk entailed in radical belief. This is so because it falls into Artworld, which makes no critique of its critique."[2] Likewise, critic Libby Lumpkin observed that this type of art "asks us to believe that theoretical intentions constitute, prima facie, political actions. . . . Art need not perform; it need only *wish* to perform."[3] Conveniently, such critiques were easily written off as being too firmly ensconced in white, Western, male-dominant aesthetics while being blind to the urgency of a political message. In this way, politically motivated art at that time became exempt from a serious dialogue with art history (by its well-meant political intentions) and from any real social consequences (by its insularity within the art world).

Still, the insistence of art with social and political aspirations (and the strength of the work by certain artists, such as Kara Walker and William Kentridge) held the prospect of an open door between art and the nonart world in ways that were as ripe with possibility as anything since Pop Art broke the hold of painterly abstraction in the 1960s. It is difficult to examine Jordan's work without thinking of such precedents. In particular, Chris Burden's projects, which utilized huge arrays of materials to make statements about power and politics, come to mind. Like Jordan, Burden did not make overt judgments in his work but relied on the bare facts to convey his messages, as in *The Reason for the Neutron Bomb* (1979), composed of fifty thousand nickels arranged in a regular grid pattern, with a wooden matchstick glued to the top of each nickel. The numbers represented the tanks positioned by the former Soviet Union along the border of Western and Eastern Europe at that time. Similarly, *The Other Vietnam Memorial* (1991) consists of twelve huge copper panels, mounted on a central pole, on which three million names are etched, symbolizing the Vietnamese killed during the U.S. involvement in Vietnam.

More recently, works by artists such as Chuck Close, Vic Muniz, and Andreas Gursky provide examples of a similar focus on the relationship between overall image and minute detail. Such works allow viewers to see in ways they wouldn't otherwise, especially in terms of the degree of detail in large photographic images. The parallels are obvious, but how is Jordan's work different, and why?

INTOLERABLE BEAUTY AND *IN KATRINA'S WAKE*

Jordan was trained as a lawyer and between 1981 and 1991 worked as a corporate litigation attorney. In 2003 he resigned from the bar to ensure that he "couldn't go back." As he describes it: "I knew early on in my law career that it was wrong for me, but I stayed for a decade because I was afraid of taking the risk of failing as an artist. Then as I approached forty, I became aware of a new and much bigger kind of fear: the fear of getting old and realizing I had not lived my life."[4]

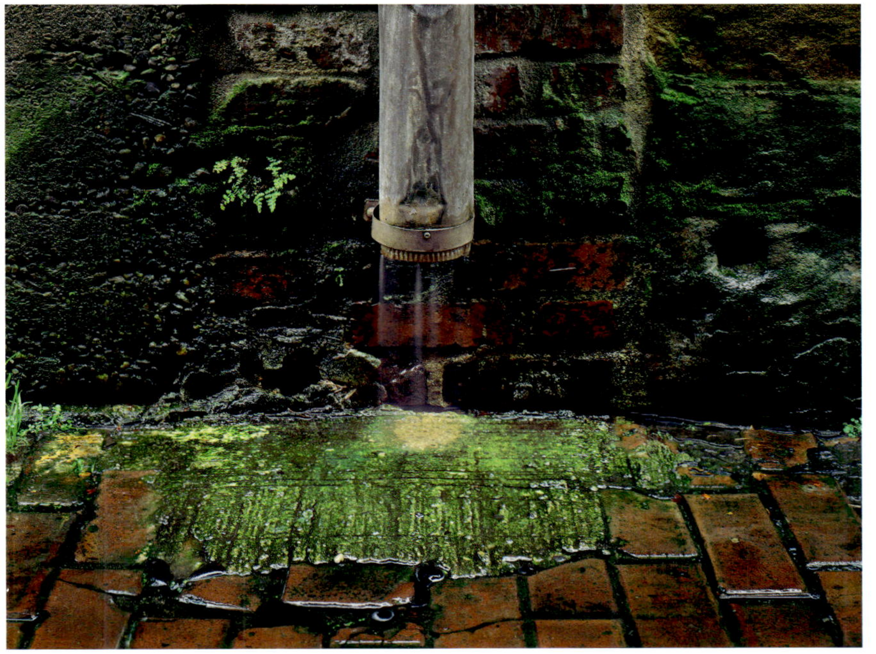

fig. 3: Downspouts and Cobblestones, 1992

During his years as an attorney, Jordan photographed as a hobby. Using 4-by-5-inch and 8-by-10-inch view cameras, he made several bodies of work that he describes as being "only aesthetic" in nature (fig. 3). His subjects varied, including Washington's forests and the alleys in downtown Seattle, but the underlying intention was to capture the complexities and beauty of color in a particular picture. "I still love some of those images," he says, "but that work was pretty disengaged. Photography for me during those years was a way of escaping rather than facing the contemporary world."

Then in 2003 he began a project called *Intolerable Beauty: Portraits of American Mass Consumption*, a series of straight documentary photographs of container sites, recycling centers, and the like (plates 1–4). These pictures have strong kinships to the traditions of landscape photography, by artists as diverse as Edward Weston, Richard Misrach, and

Edward Burtynsky. Each in his own way found beauty—and not a little pathos—in large and small evidences of the human footprint, and Jordan has extended these kinds of explorations. In his introduction to the catalogue for *Intolerable Beauty*, Jordan wrote:

> Exploring around our country's shipping ports and industrial yards, where the accumulated detritus of our consumption is exposed to view like eroded layers in the Grand Canyon, I find evidence of a kind of slow-motion apocalypse in progress. . . . I am appalled by these scenes, yet another side of me is drawn in with an entrancing sense of awe and fascination. . . . Our consumerism holds an anesthetizing kind of mob mentality: collectively we are committing a vast and unsustainable act of taking, but we each are anonymous and no one is in charge or accountable for the consequences.[5]

In 2005 Hurricane Katrina struck New Orleans, and Jordan and his wife, Victoria Sloan Jordan, made a kind of pilgrimage to respond to the disaster through his photographs and her poetry. In the resultant book, *In Katrina's Wake: Portraits of Loss from an Unnatural Disaster*, Jordan made a point of having science and environmental writers—as opposed to art writers—contribute essays. Bill McKibben wrote on hurricanes as natural phenomena: "at heart it's an expression of something purely normal; it's an equation. . . . This is how the planet is supposed to work. Or, how the planet *was* supposed to work. Until sometime in the last twenty to thirty years. . . . That's when we—meaning human beings generally and Western human beings in particular—got big. Big enough to change the equations."[6] Susan Zakin picked up from McKibben and wrote about the effects of man-made attempts to manipulate nature, noting that because of the levee system, which has sought to control the Mississippi River, "since the 1930s, the amount of sediment deposited in the delta has been reduced by eighty percent." Canals built by oil companies have further weakened the structure of marshes and thus hastened and amplified erosion. This meant that the natural wetlands of southern Louisiana could no longer serve as a strong natural buffer to the gulf waters. And so on.[7]

Jordan's photographs are poetic counterpoints to the social-scientific texts. They focus on the more intimate human stories within the disaster: the waterline on the wall of a preschool classroom, a grounded fishing boat in the middle of a neighborhood (fig. 4), purses hanging from a withered tree branch. Certainly the photographs stand on their own as poignant moments of loss, and yet the context of the essays serves to position the individual images within the big picture of environmental catastrophe.

After returning to Seattle from New Orleans, Jordan began showing the pictures from *Intolerable Beauty* around the country. "What I was hoping to get at with that project was the scale of our consumer society. The thing I noticed, after I'd been doing some talks and lectures, was that I was always using numbers. So I'd show my cell phone picture, for example, and along with the image I would say that the actual number of cell phones that are discarded in the United States yearly is more than 130 million. So this image only represents a tiny portion of that."[8] As with the Katrina pictures, these individual photographs stand on their own terms as compelling images of a disposable society; and yet they left a

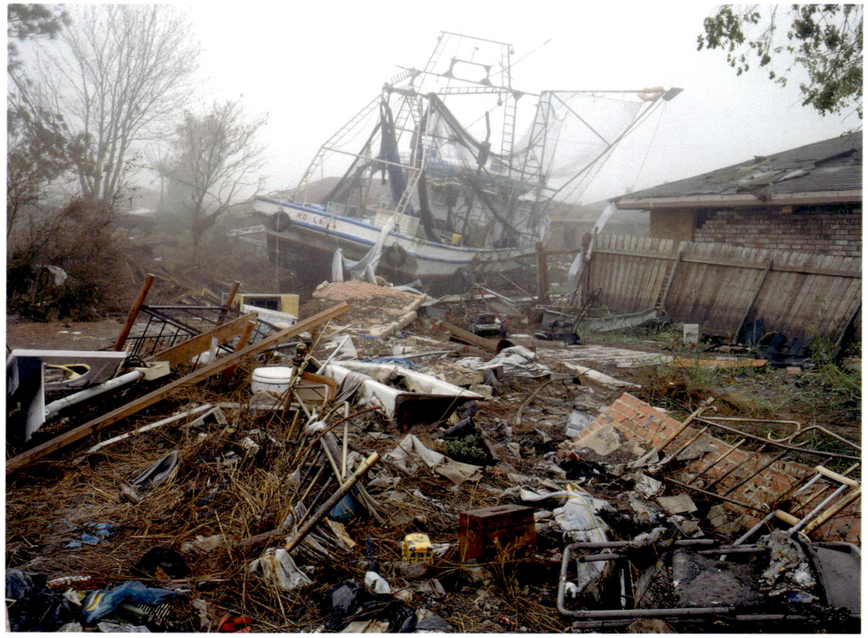

fig. 4: Fishing boat and remains of a home, Chalmette, 2005, from *In Katrina's Wake*

gap between the individual image and the overall impact he was seeking. Jordan began to realize how difficult it is for us to comprehend the vast accumulations of materials we negotiate every day, and he wanted to show "the actual quantities of things, rather than just big piles of them." However, as he has noted:

> The thing about that is, there's nowhere you can go and see these actual groups of things. It's like our entire consumer society is happening invisibly, in some ways. So the only way we know about the staggering effects we're having on our environment, for example, is to read scientific reports about statistics, but there's nowhere you can actually go and *see* the numbers. And the problem with statistics is they're so dry and emotionless. If we're going to be motivated as a culture to change our behavior, then we're going to have to find a deep motivation. Because statistics are so hard to connect with, we're not going to find motivation from them.[9]

RUNNING THE NUMBERS

Clearly, these early projects paved the way for *Running the Numbers: An American Self-Portrait*, a series Jordan began in 2005 with *Jeep Liberty*, an experiment in repeating images to convey a precise statistic (see page 15). And yet they do not explain the profound effect that *Running the Numbers* has had as imagery that visualizes our relationship to the planet. If, as art critic Dave Hickey famously wrote, "the meaning of a sign is the response to it,"[10] what is the meaning of Jordan's work as a form of communication, and what are the components of its use?

Though Jordan shows his photographs in museums and galleries and sells them as art objects, the content of and reactions to the work live predominantly outside the institutional control of the art world. To date, most people have come to know of his work by "word-of-Internet." Its validation has come through the process of one person passing it on to another. I first saw it in an e-mail sent by my sister-in-law Diane. Through such private and democratic forms of circulation, Jordan's art has reached a worldwide audience. The artist's Web site receives an average of 75,000 individual visitors per month (more than *Art in America*'s 67,300 annual subscribers). Jordan himself has appeared on television shows such as *Bill Moyers Journal*, *The Colbert Report*, and the *Rachel Ray Show*, and in several documentary films. He presents illustrated lectures at such diverse gatherings as the TED (Technology, Entertainment, Design) Conference, the Greener Gadgets Conference in New York, the Mountainfilm Festival in Telluride, and cultural outreach programs such as one recently hosted by the mayor of Caracas, Venezuela. Jordan's art is, by any measure, widely received and has a popular appeal.

His timing has been fortuitous: even five or six years ago, the Internet was in a much slower stage of development. Today he can count on broadband access in much of the world. One of the things the Internet does well is to personalize the relationship between content and user. The mouse click seems to enhance our sense of personal discovery in a way that is different from a visit to the more public space of a museum or gallery. The Internet also encourages sharing, whether it is a simple e-mail joke, a YouTube offering, or a piece of music. Jordan's work, being digital from inception and minimalist in form, communicates effectively online. Because the work is received in such an extensive and transcategorical manner, it not only positions itself at a prime intersection between aesthetics and social conscience but also implicitly positions other works of art as arbitrarily and even unnecessarily cloistered in the art world. Why is it that this work reaches so many people? What need does it fulfill?

Wired magazine editor Daniel Pink suggests an answer in his recent book *A Whole New Mind: Why Right-Brainers Will Rule the Future*. In it, he proposes that we need "pattern recognizers and meaning makers."[11] As he explains:

> Today facts are ubiquitous, nearly free, and available at the speed of light. . . . An English-speaking thirteen-year-old in Zaire who's connected to the Internet can find [current information] . . . as quickly and as easily as the head librarian at Cambridge University. . . . When facts become so widely available and instantly accessible, each one becomes less valuable. What begins to matter more is the ability to place facts in *context* and to deliver them with *emotional impact*.[12]

Pink goes on to say that our "well-developed high-tech abilities" need to be supplemented with "high-concept and high-touch" skills, such as the ability to create artistic and emotional beauty; to detect patterns and opportunities; to craft a satisfying narrative; to combine seemingly unrelated ideas into a novel invention; and to stretch beyond the quotidian, in pursuit of purpose and meaning.[13]

Jordan's work responds to this challenge, and it does so with the clear goal of connecting the individual to collective issues. He says:

> My idea with the *Running the Numbers* series is to . . . give you the statistic in a different way that allows the viewer to experience the number more directly with their heart. One of the huge problems that faces our society right now is that . . . our consumerism, and the resulting global warming, worldwide environmental destruction, the toxification of our oceans, and the desertification of our agricultural lands, and so on, are not happening because there's an extremely bad person out there who is doing a huge amount of terrible consuming. This is happening because of the tiny incremental harm that every single one of us is doing as an individual. The problem is this cumulative effect from the behaviors of hundreds of millions of individuals. Each person looks around at his or her own behavior, and it doesn't look all that bad. What we each have to expand our consciousness to hold is that the cumulative effect of hundreds of millions of individual consumer decisions is causing the worldwide destruction of our environment.[14]

There is a terrible question at the heart of these works, for as we drill down to the details and then back out to overall issues, we cannot help but encounter the essential dilemma regarding the role of the individual within such overwhelming problems. As much as we like to believe that "every vote counts," the real effect of one person's actions in relationship to six billion others is all but invisible. What if I *do* stop using plastic bags? What if I *don't*? Psychologically speaking, how do we convince ourselves that one person matters?

This is precisely where Jordan's work hits the mark: it positions us to confront this dilemma of "the one in the many." Seeing the real object shifts the viewer's experience to another level based on a dynamic physical and perceptual relationship. The imposing, larger-than-human scale implicates viewers in the scene, positioning us to "take a stand" in relation to the subject. The scale also enhances our active perceptual reconstruction of the image—from far to near, from broad to intimate—and reinforces the act of scrutiny. The experience is similar to viewing a pointillist painting in which thousands of parts combine to form an overall image, and yet, on closer examination, the collective effect explodes into tiny details, rendered individually, one at a time. In addition, Jordan's use of vanishing-point perspective extends the active engagement of the viewer even further. Rather than merely conveying statistics, which could easily be done through *USA Today*–type graphs, he creates pictures that pull us in and offer spaces for contemplation, often invoking traditions of landscape art that are steeped in the iconography of "beauty": for example, Impressionism (*Cans Seurat*), Japanese prints (*Paper Bags*; plate 11), and Sierra Club calendars (*Denali Denial*; plate 5).

fig. 5: Lightbulbs, 2008 (detail of plate 20)

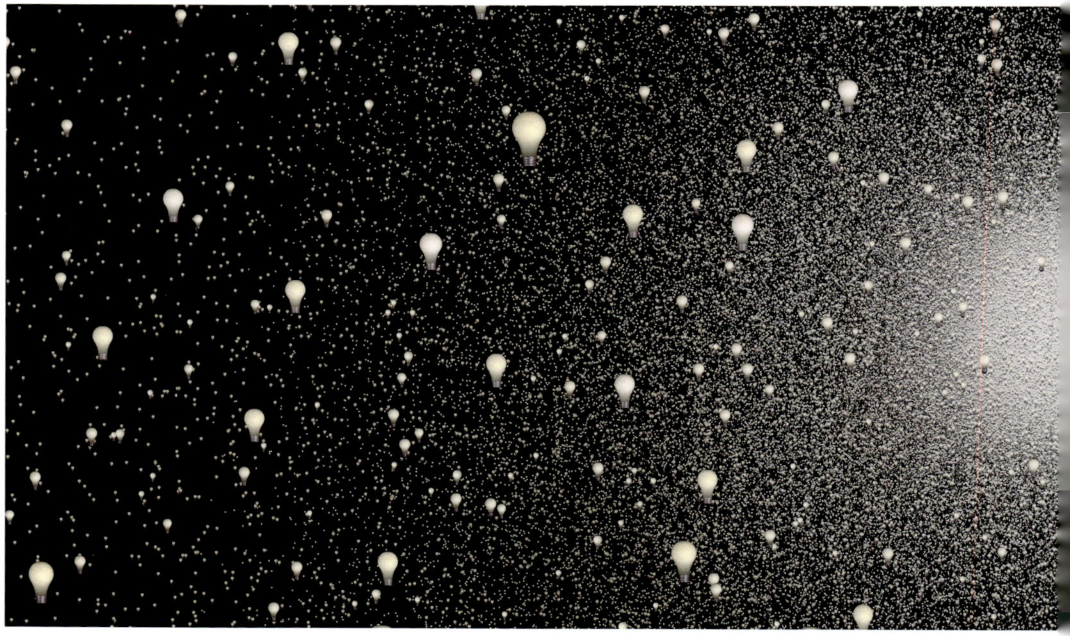

The use of the Seurat image certainly suggests a reference to art as a constructed scene that evokes only the most pleasant associations. Similarly, *Toothpicks*, with its infinite sky and mirrored crop rows, evokes an all-American pastoral ideal akin to a Grant Wood landscape or the opening sequence of the musical *Oklahoma!* (plate 18). *Lightbulbs* extends the prospect of space to the wonder and drama of the "final frontier" (plate 20; fig. 5). By contrast, the use of vanishing-point perspective in *Plastic Bottles* intensifies our awareness of what has been lost to our consumptive practices: this is not just a collection of "stuff" but the vision of a vast, post-apocalyptic wasteland.

Jordan's work, which at first glance is so transparent in terms of presenting "just the facts," poses uneasy questions that are at once personal and global. His art transcends the normal gap between art and society, and is finding a place within an ever-expanding number of social and environmental groups all over the world, each one acting as a small part of a vast, bottom-up process, which, when taken together, form what environmentalist and author Paul Hawken refers to as the "largest social movement in all of human history." He goes on to describe what he sees as "ordinary and some not-so-ordinary individuals willing to confront despair, power, and incalculable odds in an attempt to restore some semblance of grace, justice, and beauty to the world."[15] The success of Jordan's project is that it does not stand outside the fray and issue attacks but is made by a coparticipant who is grappling to understand his own place within a wide and diverse audience, both as an artist and as a world citizen.

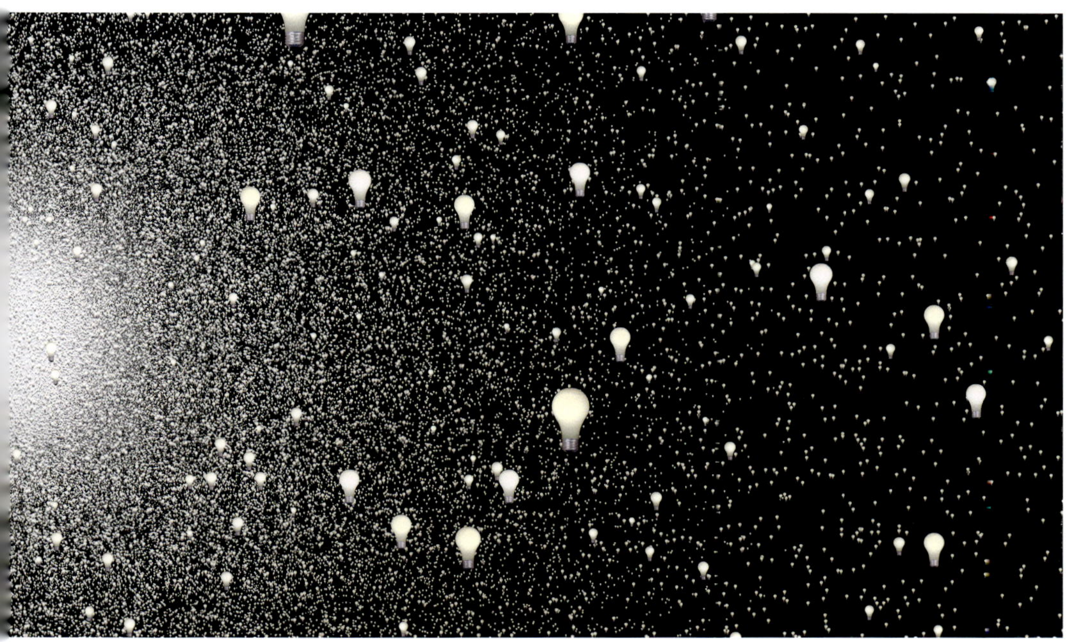

Chris Bruce is director of the Museum of Art/Washington State University.

NOTES

1. Gary Trudeau, *Doonesbury*, Universal Press Syndicate, 2007.
2. Robert Hughes, *American Visions: The Epic History of Art in America* (New York: Alfred A. Knopf, 1997), 613.
3. Libby Lumpkin, *Deep Design: Nine Little Art Histories* (Los Angeles: Art Issues Press, 1999), 93.
4. Unless otherwise noted, statements by the artist are from conversations with the author in 2007 and 2008.
5. Chris Jordan, *Intolerable Beauty: Portraits of American Mass Consumption* (Seattle: self-published, 2005), 3.
6. Bill McKibben, "Year One of the Next Earth," in Chris Jordan, *In Katrina's Wake: Portraits of Loss from an Unnatural Disaster* (New York: Princeton Architectural Press, 2006), 9–10.
7. Susan Zakin, "A Corner of the Fallen World," in Jordan, *In Katrina's Wake*, 57.
8. "Running the Numbers," Chris Jordan interviewed by Nicole Pasulka, *The Morning News*, New York, July 23, 2007, www.themorningnews.org.
9. Ibid. Jordan describes how he gathers his data: "I do my own research on the Internet. Usually with a few searches I can find government databases (such as the enormous Centers for Disease Control Web site) and Web sites that point to other references and so on. When I find a few different sources that all agree on a particular number (say, a *New York Times* article, a reference in *Harper's*, and a politically neutral Web site), then I feel comfortable that I have a defensible number to use. Whenever there is a question about a particular statistic, I take a conservative approach. I keep all the articles and references I find, so if there is ever a question about one of the numbers, I can provide backup." E-mail to the author, July 3, 2007.
10. Dave Hickey, *The Invisible Dragon: Four Essays on Beauty* (Los Angeles: Art Issues Press, 1993), 29.
11. Daniel H. Pink, *A Whole New Mind: Why Right-Brainers Will Rule the Future* (New York: Riverhead Books, 2006), 50.
12. Ibid., 102–3.
13. Ibid., 51–52.
14. Pasulka interview, *The Morning News*.
15. Paul Hawken, *Blessed Unrest: How the Largest Movement in the World Came into Being and Why No One Saw It Coming* (New York: Viking, 2007), 4.

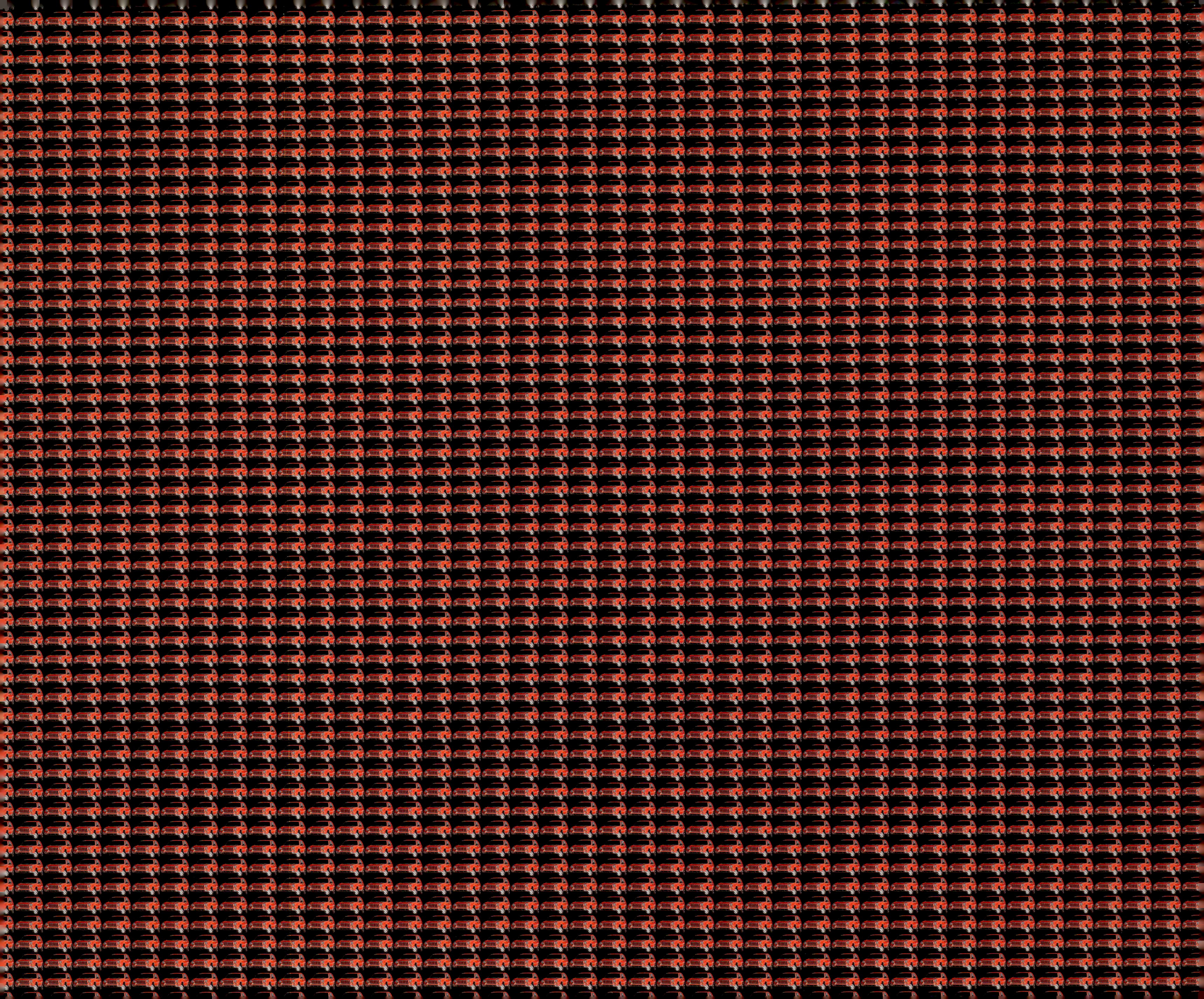

Lucy R. Lippard

RUNNING (AWAY FROM) THE NUMBERS

How come nobody thought of this before? Like all significant activist art, Chris Jordan's *Running the Numbers: An American Self-Portrait* fuses image and information into one big double whammy (I see the series as a "POW!!" in explosive comic-book style). Like all brilliant ideas, it's relatively simple: photograph hundreds or thousands of objects such as tossed-out plastic bottles or office paper or SUVs, then digitally reproduce them to represent thousands or millions, and then title the huge, perversely beautiful, almost abstract images with numbers that literally bring them down to earth and into your face/my face/our face.

Just when photography has lost any pretensions to truth (a long process ranging from Edward Curtis to Photoshop), Jordan rides to the rescue with a new way to tell it like it is. His images offer us what we should already know but conveniently ignore. Plastic bottles? Two million of them, the number used in the United States every five minutes (plate 6). Office paper? The 30,000 reams or 15 million sheets equal the amount used in this country every five minutes. SUVs? *Jeep Liberty* depicts 200,000 of them, the number sold in the United States every year (fig. 1). (This figure has, I hope, declined since the rise of oil prices and the subsequent economic debacle, though I heard the other day on the radio that Hummers no longer desired by the average American are being shipped to China and India, where they are still popular among the greedy classes.) *Jeep Liberty* was in a show on climate change I curated in 2007 in Prius-proud Boulder, which turned out to harbor more SUVs than any other city in Colorado.

fig. 1: Jeep Liberty, 2005 (detail at left)

The first work of Jordan's I saw (*Denali Denial*, at the Lannan Foundation, Santa Fe, in 2006) seemed at first glance to be a big pixelated black-and-white photograph of the gigantic Alaskan mountain and its melting glaciers (plate 5). As I approached, I found that the image was constituted of the word "denial" and the logo of the GMC Yukon Denali. (Hmmm. Ansel Adams this is not. . . .) When I read the extended caption, I was informed that the photograph's 24,000 logos represented six weeks of sales of the SUV in 2004. I had been transported in a matter of minutes from admiring a mountain to deploring a lifestyle.

Jordan calls his formal strategy "hypermodernism," a kind of Google Earth phenomenon, in which he says he uses "very large numbers of small things to create images of extreme visual complexity. They speak to the relationship of the individual and the collective, and also the relationship between the near and the far."[1] He has come up with a new way to make statistics stick. I can't repeat the numbers but I always remember that they are very, very large and bode no good. The shock sparks a metaphor for a society that rarely looks or gets a chance to look at anything up close or in contemplative time. The TV news is whisked past us by a mealy-mouthed "anchorperson." We dash through landscapes without being told what they contain and what has been done to them. (This is where the Los Angeles–based Center for Land Use Interpretation, or CLUI, so valuably weighs in with its data bank.) Education is demoted to test-passing, sports to virtual athleticism. Even democracy has bypassed us, as Rebecca Solnit remarks: "By the time the political process has reached the voting booth, all the real choices have been programmed in, and the voter becomes a consumer."[2] We are entertained by simulation of one kind or another, some of which offers the illusion of participation. In Jordan's work we get the macro- and the microcosm interwoven. One after the other, and back again. There's no escape.

This strategy of mind-boggling multiplicity actually has a long activist art history in and out of the more or less mainstream art world (I think of a Colorado couple's ceramic miniaturization of our nuclear arsenal, Ann Hamilton's labor-intensive installations, and Steven Deo's globe created from thousands of plastic toy soldiers, among many other examples in the last few decades). Nor is Jordan's use of multiplicity unprecedented in photography (Ed Ruscha, Robert Flick, Edward Burtynsky, and others have been there before), but he transforms this formal device into an arrow to the heart of much that is wrong with the first world in the twenty-first century. Of course his stunning images are not as collectible as they might be without the captions. Seen from a certain distance, and untitled, they are impressively minimal, formal, vague, and contentless. I wonder if collectors are buying them and leaving off the captions, a common occurrence for politically charged art that is of course extremely frustrating for the artists, who usually need the money and are glad to make a sale but in the process lose control over their own work.

Jordan's figurative images—*Cans Seurat* (*A Sunday on La Grande Jatte* re-created with 106,000 of the aluminum cans that litter American roads and picnic grounds, equaling thirty seconds of national consumption), breasts made up of 32,000 naked Barbie dolls (the number of elective breast augmentation surgeries performed monthly in the United States in 2006), and a smoking skull, derived from a Van Gogh painting, made from 200,000 cigarette packages (the number of Americans who die from smoking every six months)—are skillful and eye-catching (plates 10, 15, 17; figs. 2, 3). They certainly make their point in a directly populist manner, like the anonymous portrait of George W. Bush made from the faces of American dead in Iraq. But the giant fields of waste, of crushed cars, of cell phones, of prison uniforms, once reunited with their statistical significance, are even more powerful social commentaries. The time it takes to move from expansive visual intricacy to close-up specificity is momentous.

Most recently, the hyperenumerative strategy has been employed (by Jane Hammond, Athena Tacha, and Isabella Gonzales, among others) to communicate the statistical realities of the body counts—civilians and U.S. soldiers—in Iraq and Afghanistan. Can plastic bottles and SUVs be compared to the victims of an impeachable conflict collaged together from decontextualized and faked documents? No, but at the heart of both is *waste*, and waste is what Jordan is deploring. The waste of human lives and the waste of natural resources that will end up by killing people are not so different in the long run. Jordan has chosen not to attack

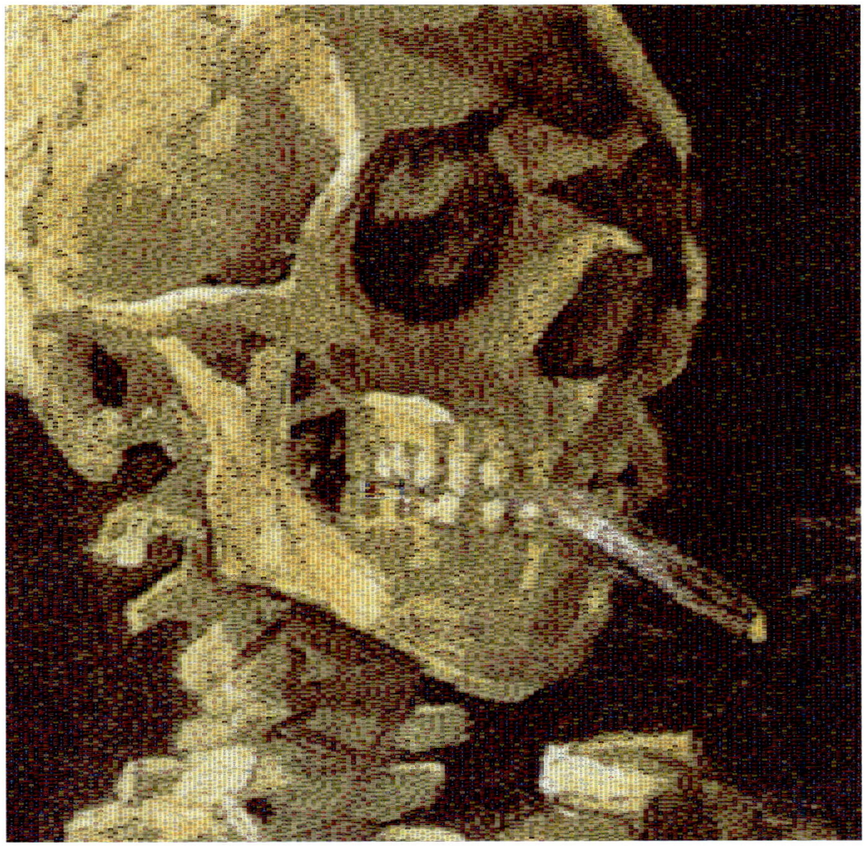

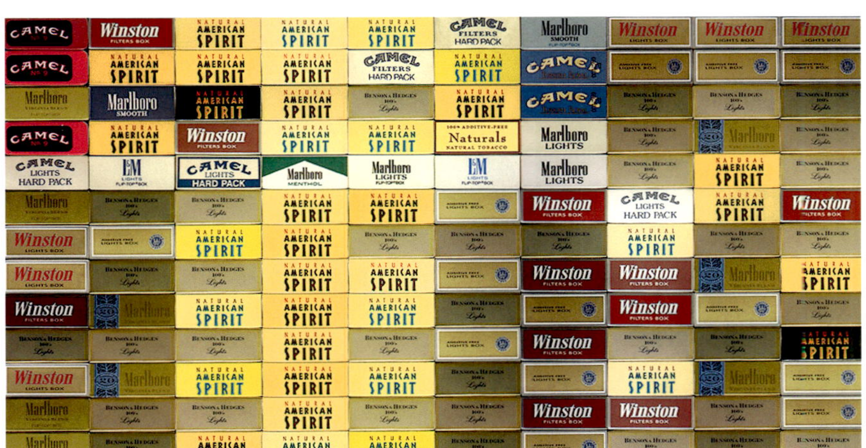

figs. 2, 3: Skull with Cigarette, 2008 (details of plate 15)

directly the powers that be, the powers that profit from these conflicts, but to lay the responsibility at the feet of us, the actual consumers, who fall for gimmicks like bottled water, gas-guzzling vehicles, and "patriotic" wars. Most of us pay the taxes necessary to pump up the Department of "Defense." We can't avoid it without great disruption to our lives, which few can afford. But the fact remains that carrying a cloth bag to the supermarket and using personal water bottles involve little moral fortitude and hardly damage our touted national "quality of life." *Waste not, want not*. Who knew the ramifications of forgetting that thrifty advice, which I heard so often in my Depression childhood.

> All earth's productive land and sea, divided equally among its population of six billion, leaves only 4.7 acres to support each person . . . [but] the average American's lifestyle demands 30.9 acres. This suggests that for the whole world to enjoy America's standard of living would require another five earths. In fact, if America used more renewable resources (particularly for energy), recycled more, and introduced the far more efficient means of production and distribution already available, or readily developed, its footprint could be reduced to a fraction of its present size without compromise to its standard of living.[3]

This past summer NPR ran a series of street interviews asking "What is an American?" The all-too-predictable responses ranged from patriotic cliché to self-congratulation to idiocy to (occasionally) introspection. No one said what I was waiting to hear: An American is an overprivileged, underaware person with the attention span and immediate gratification demands of a four-year-old, who is on her/his way to ruining the world that has provided us with so much so far. My mantra these days—*long-term thinking is in short supply*—is, alas, particularly appropriate for this country where we live so happily out of the way of the genocide, famine, and natural disasters that plague the rest of the world. (Though it's true that September 11 and Katrina hit us hard, the numbers pale beside those killed, imprisoned, and made homeless by tsunamis, earthquakes, wars, and economic exploitation elsewhere.)

The austerity of the means by which Jordan takes the measure of our society—boring statistics, visual repetition—only adds to the impact. The works have to be large (often very large, as in the 10-by-25-foot *Valve Caps*) because the numbers are so large: 3.6 million valve caps representing one year of SUV sales in the United States. Jordan's statistics alone are political poetry, just as the photographs offer their own visual poetics. Toothpicks? One hundred million of them equal the number of trees harvested in the United States every year to make paper for mail-order catalogues (plate 18). Supermarket paper bags? We consume 1.14 million every hour (plate 11). Jet trails? Eleven thousand of them, the number of commercial jet flights in the United States every eight hours (plate 12; fig. 4). And so it goes.

Once consumerism was just consumerism, disgraceful enough, but shrugged off as a social peccadillo shared by most Americans. Now it is inextricably entwined with global warming. Every one of those sheets of paper, plastic bottles, and SUVs bloats the carbon footprint that should be shrinking. But consumption is so convenient. Nobody's pure. Nobody's innocent. But if most of us were trying, what a different world it would be. George W. Bush's suggestion after September 11 that the nation meet the crisis by shopping was predictably as far as that small mind could go. (And I guess the thrift shops where I get everything I need benefit from everybody else's consumerism while satisfying mine.) It would be so easy for recycling to become an all-pervasive way of life, as yard-sale aficionados already know. And what happened to "normal"-sized houses? Why does a middle-class family think it *needs* a monstrous Chevy Tahoe and a 10,000-plus-square-foot house? Why do the rich *need* several even larger "homes"? Why do the superrich feel entitled to a supersized footprint? Is the economy dependent on this gigantism? Well, imagine a new kind of economy that is constructive rather than destructive.

And while we're at it, imagine a new kind of art world in which creative production and practice are not limited to museums (necessary for preservation and public viewing) and galleries (artists have to live on something) but are as ubiquitous as commercials. Imagine art schools

fig. 4: Jet Trails, 2007 (detail of plate 12)

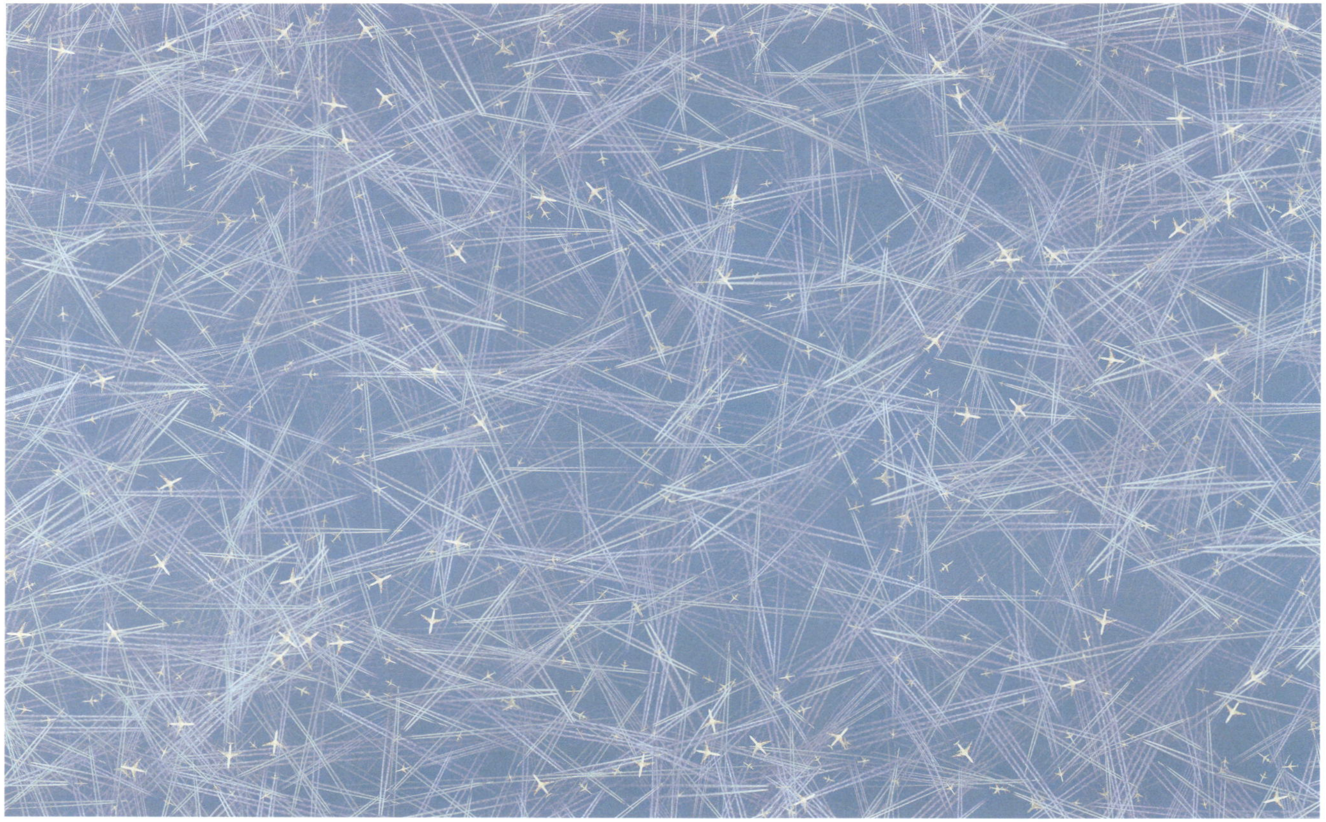

in which students are trained to work *for* something and *with* someone. Imagine a national guaranteed minimum wage that allows more people to be artists and to acquire art, an art world where academic jargon is not the coin of the realm, which is run not by money but by vision. Don't get me started.

I've spent forty years conspiring with artists who either want to stay in the art world and gain some control over how their work is handled or want to get out of the art world, reach a broader audience, "make a difference," and—what the hell—change the world. Many of them are criticized from within the art world for making work that is too obvious (that is, too close to the mark), too "politically correct" (god forbid! Is the alternative really preferable?). They are criticized from without when they resort to aestheticization of the dangerous and evil, or to irony. ("The people's voice is not ironic," I pontificated years ago, which is not to say "the people" are too dumb to get it, but that they tend to perceive "high" art as aimed literally over their heads.) Jordan, who worked for a decade as a lawyer, has earned opportunities to make his case to the world at large, to nonart groups, to wide TV audiences. Because his case is our case, this is decidedly to everyone's advantage.

Among the ideas that have come up in art activist groups over the decades is to make all museums include artists on their boards of trustees (and just to be sure they don't choose ringers, every artist in the constituency would vote for three names; the top of the list, bound to be well known, would be the pool from which the museums could choose). This was a brainstorm from the 1960s (maybe you could tell?), when institutional critique took place in the streets as well as in the journals. Decentralized museums was another suggestion/demand: works from major museum collections (which hold more art in storage than is ever shown) would be doled out to community centers in different cultural neighborhoods; they would choose what they wanted. Today we have the global Guggenheim, Dia Beacon in a former factory up the Hudson River, huge earthworks in the western United States that are or soon will be open to the public, but not yet a decentralized distribution system for physical artworks.

However, we have the Internet, and as Chris Bruce details in his essay in this volume, Jordan's work, impressive as it can be in its physical manifestation, is a natural for cyberspace. At this point, it becomes "conceptual art," and the system that sends it around the world is heir to those exhibitions in the 1960s that could be transported in a single box or suitcase, or not transported at all, as in Seth Siegelaub's *March 1–31, 1969*, which featured a different artist each day, making work that took place all over the world, brought together only in the catalogue. Hans Haacke and the N. E. Thing Co. (Iain and Ingrid Baxter) predicted aspects of the World Wide Web with their Telex pieces. The Internet, that inconceivable tangle of neurons, all too reminiscent of Alzheimer's disease, is a foreign land and a foreign language to a Luddite like me. But it is exactly what we were wishing for, forty years ago.

Someone said that artists are valuable because they can work between the questions and the answers. Jordan works with answers we do not want to hear, although it is more likely that nonartists will come up with the solutions. Maybe some artist will discover a way to counter the bad news with some good, extolling solar panels and low-flow toilets and green buildings. Too simple? Too "politically correct"? In activist art, it is always more difficult to support something than to oppose something. Meanwhile, "pessimism of the intellect" coexists with "optimism of the will," as the political theorist Antonio Gramsci said. The advent of Chris Jordan's work and that of other artists toiling with or without recognition, within and outside of the art world, especially the hundreds of small collectives working all over the world across national boundaries and political taboos—this is what gives me hope.

Lucy R. Lippard is a writer and activist, author of twenty books on contemporary art and cultural criticism, including, most recently, On the Beaten Track: Tourism, Art, and Place.

NOTES

1. Chris Jordan, personal communication with the author, June 9, 2008.
2. Rebecca Solnit, "The Garden of Merging Paths," in *Storming the Gates of Paradise: Landscapes for Politics* (Berkeley: University of California Press, 2007), 61.
3. Peter Buchanan, "Invitation to the Dance: Sustainability and the Expanded Real of Design," in William S. Saunders, ed., *Nature, Landscape and Building for Sustainability: A Harvard Design Magazine Reader* (Minneapolis and London: University of Minnesota Press, 2008), 116.

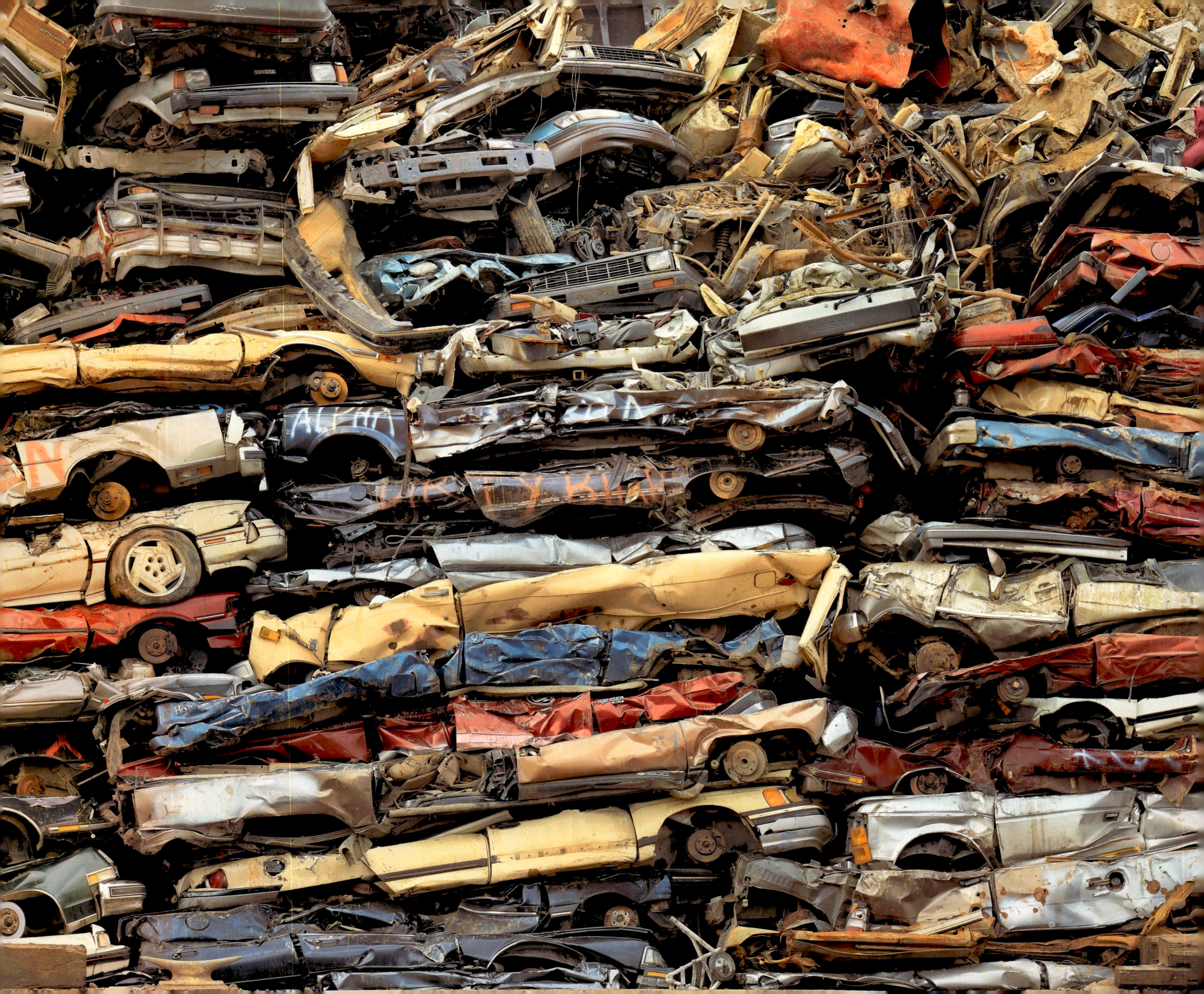

INTOLERABLE BEAUTY
PORTRAITS OF AMERICAN MASS CONSUMPTION

Exploring around our country's shipping ports and industrial yards, where the accumulated detritus of our consumption is exposed to view like eroded layers in the Grand Canyon, I find evidence of a slow-motion apocalypse in progress. I am appalled by these scenes, and yet also drawn into them with awe and fascination. The immense scale of our consumption can appear desolate, macabre, oddly comical and ironic, and even darkly beautiful; for me the consistent feature is a staggering complexity.

The pervasiveness of our consumerism holds a seductive kind of mob mentality. Collectively we are committing a vast and unsustainable act of taking, but as individuals we each are anonymous and no one is in charge of or accountable for the consequences. I fear that in this process we are doing irreparable harm to our planet, our culture, and our individual spirits.

As an American consumer myself, I am in no position to finger wag; but I do know that when we reflect on a difficult question in the absence of an answer, our attention can turn inward, and in that space may exist the possibility of some evolution of thought or action. So my hope is that these photographs can serve as portals to a kind of cultural self-inquiry. It may not be the most comfortable terrain, but I have heard it said that in risking self-awareness, at least we know that we are awake.

~cj (2005)

Crushed Cars #1, Tacoma, 2004 (detail)

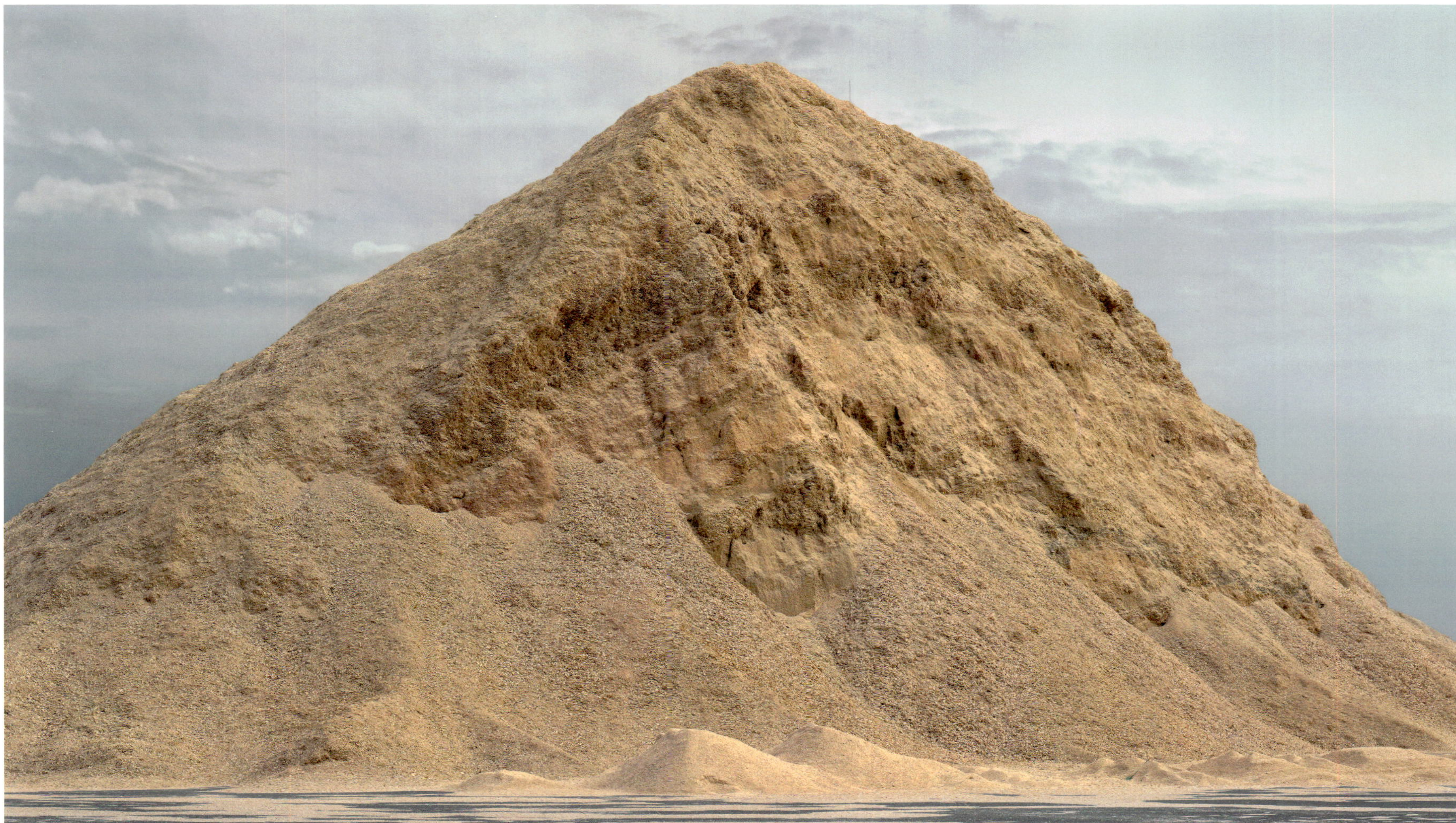

(plate 1) Sawdust, Tacoma, 2004

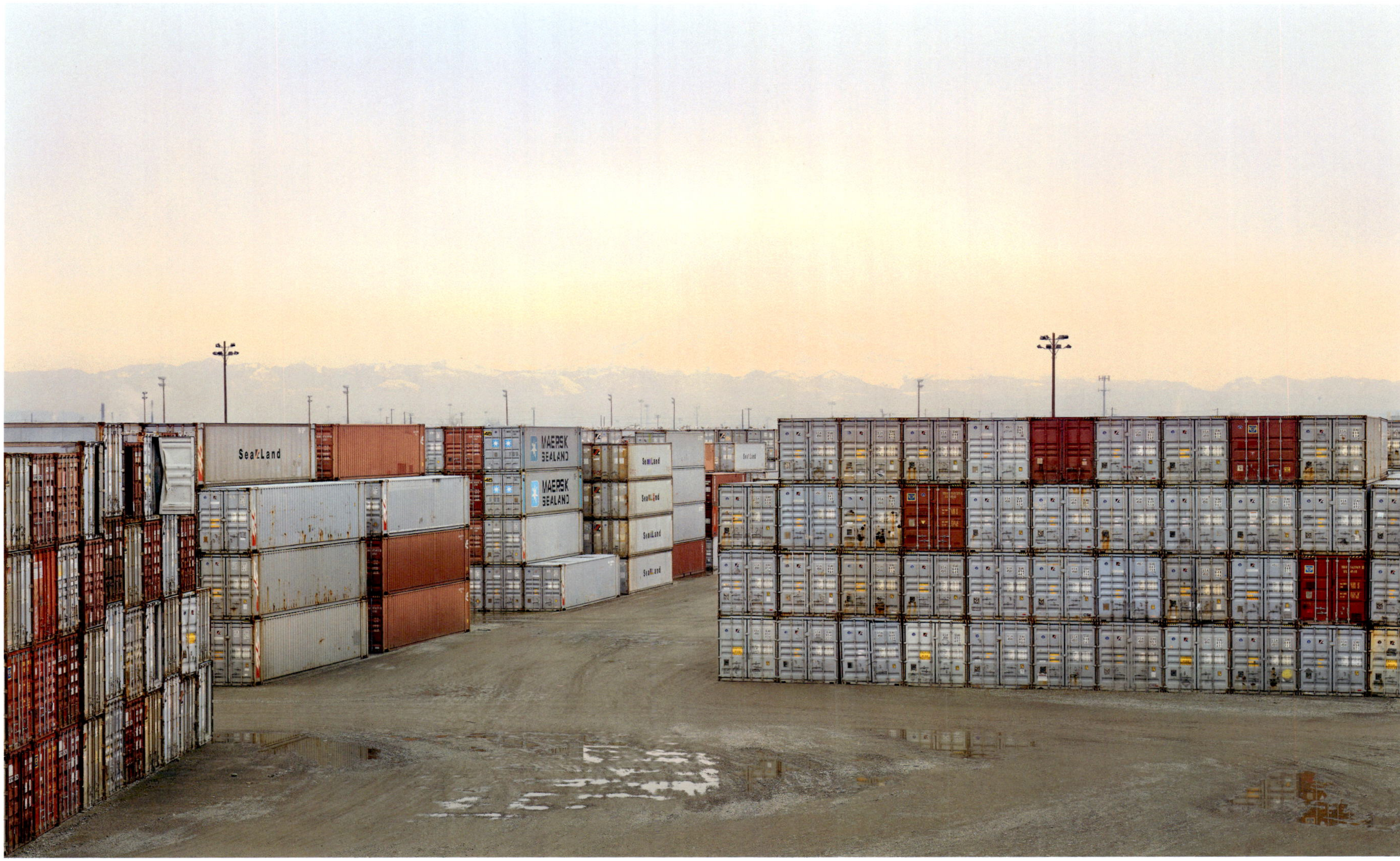

(plate 2) Containers and Mt. Rainier, Tacoma, 2004

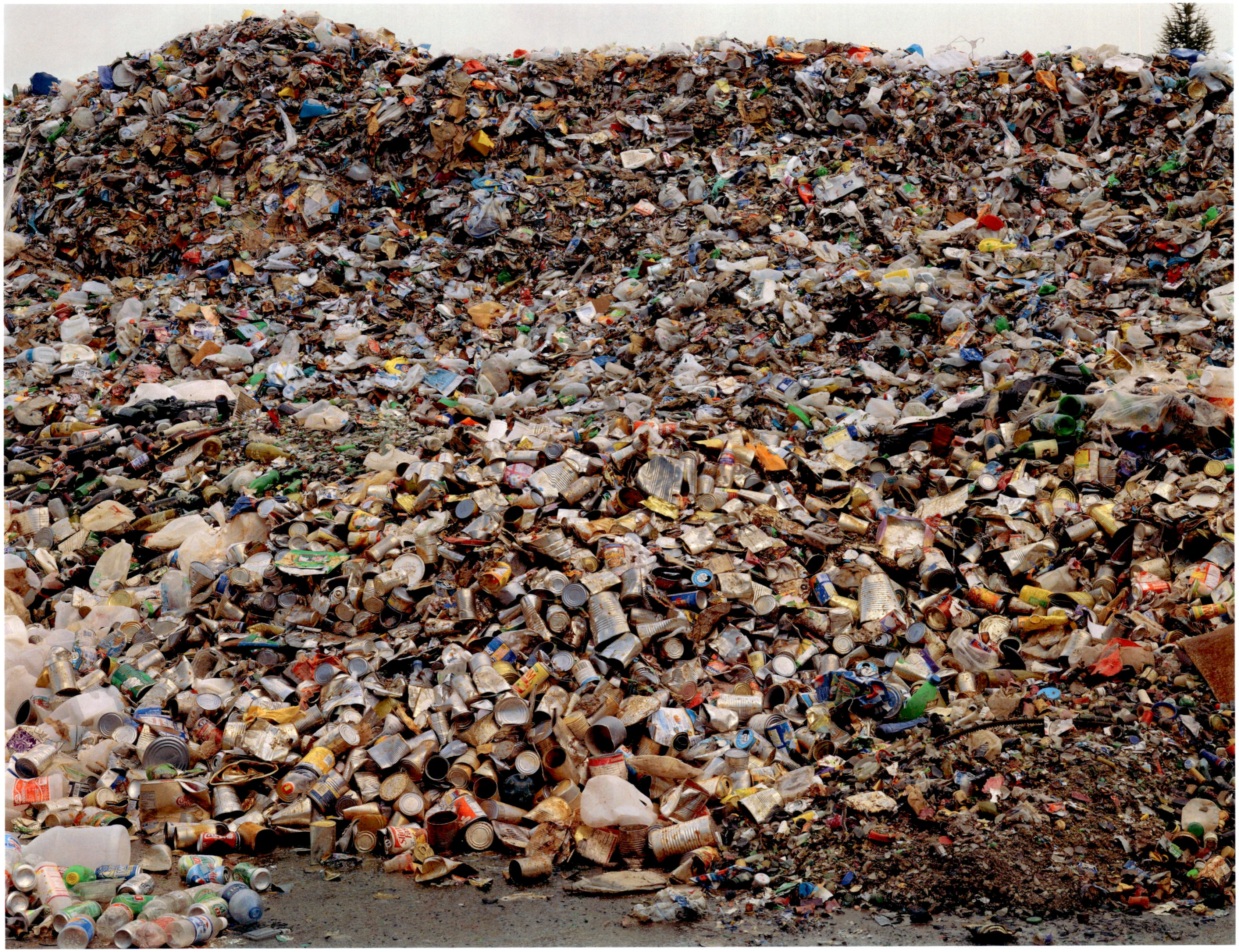

(plate 3) Mixed Recycling, Seattle, 2004

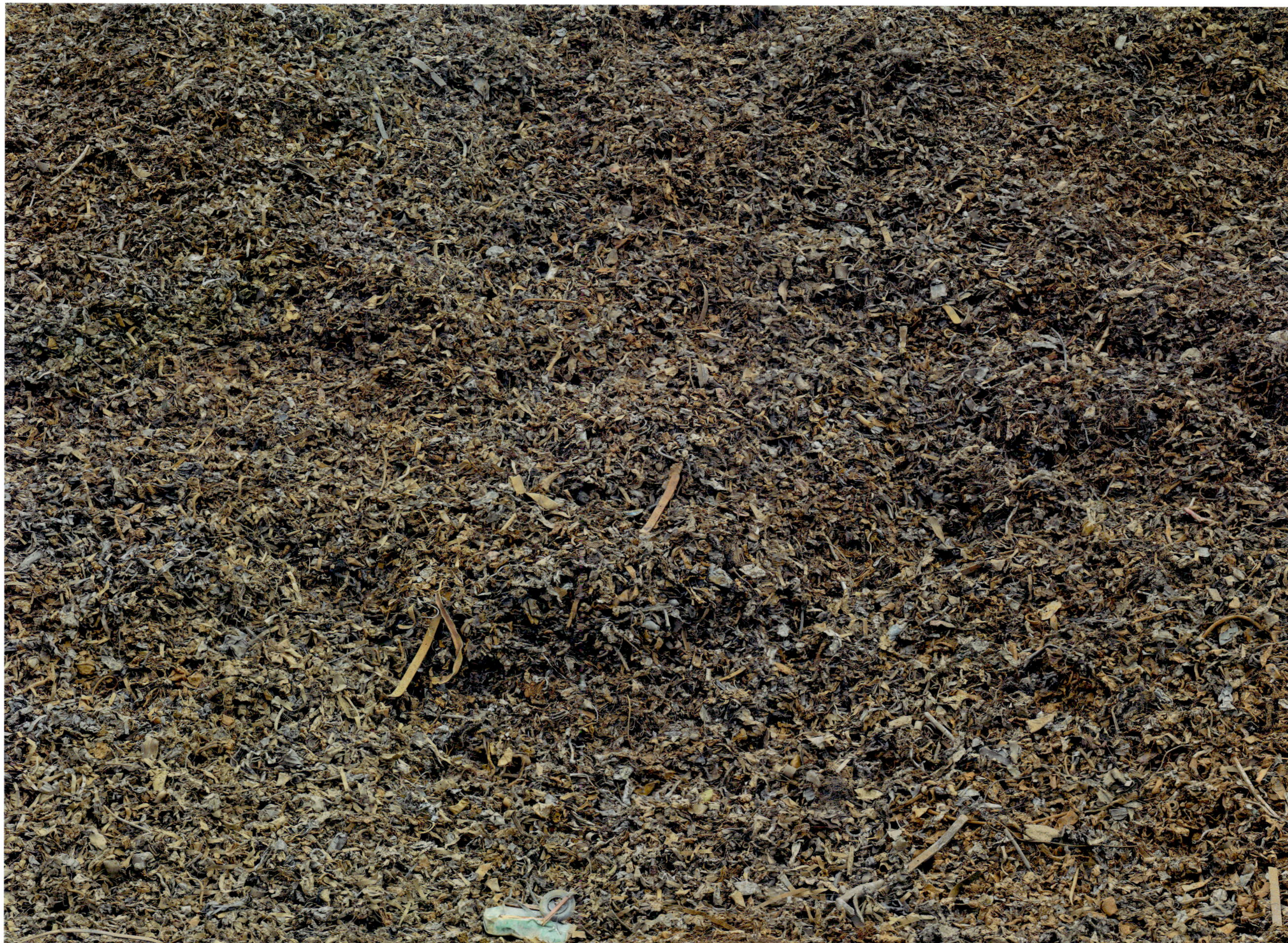

(plate 4) Steel Shred, Tacoma, 2004

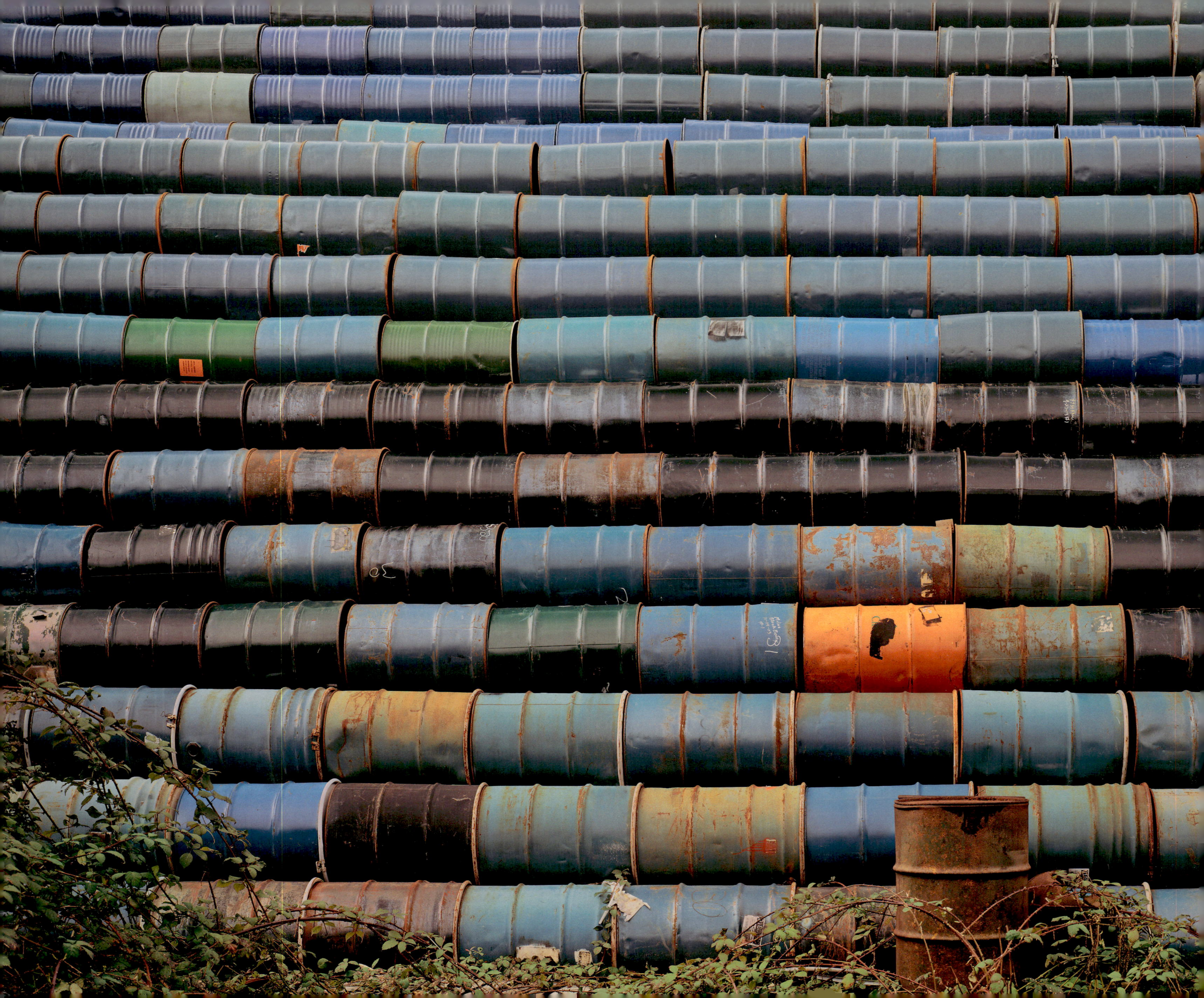

Paul Hawken

THE IDYLLIC AND THE UNFORGIVABLE

A wasteful society is a relatively new phenomenon. I spent part of my childhood on a farm belonging to my Swedish grandmother and Scottish grandfather, and very little was thrown away. The barn was full of used washers, bolts, wire, and doodads. In the kitchen, every other plate was chipped, but the china never left the dinner table with food on it. Gravy and juices were mopped up with homemade bread, vegetable peelings went to the chickens, the eggs were eaten at breakfast, the eggshells were put into the coffee grounds, the coffee grounds were placed in the compost, the compost was tilled into the garden, the tomatoes and corn from the garden were sealed in glass jars that joined the jams and jellies lining cool, dark basement walls. Paper lunch bags were brought back from school and neatly folded for the next day. Our idea of play included capturing horned toads and pretending they were dinosaurs or lying face up in irrigation ditches, the tiny fry tickling our toes, and imagining we were floating down a great Amazonian river. Our notion of a toy was a bald tire swinging from the sycamore. Had my grandparents been from Chile, Korea, or Kerala, it essentially would have been the same. Nothing would have been wasted. That is not true today.

Waste is food for life or the source of death, depending on whether it is made by living systems or industrial ones. Industrial waste is the unseemly shadow of our gross domestic product, the unwanted ghost of manufacturing that will visit us far into the future, haunting our rivers, soils, and bloodstreams. The drinks contained in plastic bottles are consumed in minutes, but the polymers last for centuries. Commercial carpets are discarded after eight years, but their PVC backing will remain in landfills for 300 centuries. Plutonium waste from nuclear power plants will need to be vigilantly guarded for 2,500 centuries (civilization is about 70 centuries old). We are newbies on the planet and so is our type of waste. Much of the waste humans make is something the planet has never seen before, chemicals and compounds that are put together with such force and manipulation they cannot easily be taken apart. This new form of offal piles up, pollutes creatures, poisons waterways, and strangles marine life. For Chris Jordan, our throwaway life is a telltale handprint of civilization in trouble.

The natural world produces more waste than the debris created by the 6.7 billion people who inhabit the earth, but in nature everything that decays, crumbles, or rots becomes food for invisible detritivores munching in the fecund darkness of soils and oceans. The White Cliffs of Dover are layers of calcium carbonate built up over tens of millions of years, a boneyard of coral, sponges, and small creatures that started piling up in the tropical waters of the Cretaceous period. Much of our industrial feedstock in minerals was the excreta of bacteria deposited eons ago. The food we eat depends on the decomposition of the soil beneath it. Life flourishes because waste is nutritious and transient. Nature literally goes to seed. This is a good thing. Nature's waste does not become carcinogenic when ingested. It does not insert mutagens into pregnant women. It does not concentrate toxins in rusting barrels stored in salt caves.

Our waste is such an unconscious part of existence that it passes little noticed. Visibly, there are waste bins, landfills, and trash-strewn highways.

fig. 1: Drums, Seattle, 2004

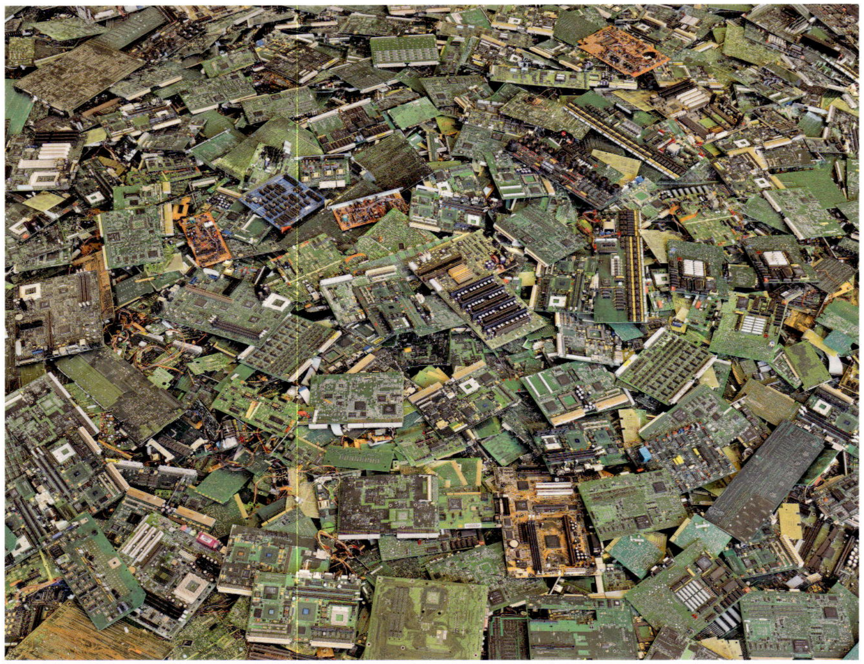

fig. 2: Circuit Boards, Atlanta, 2004 (detail)

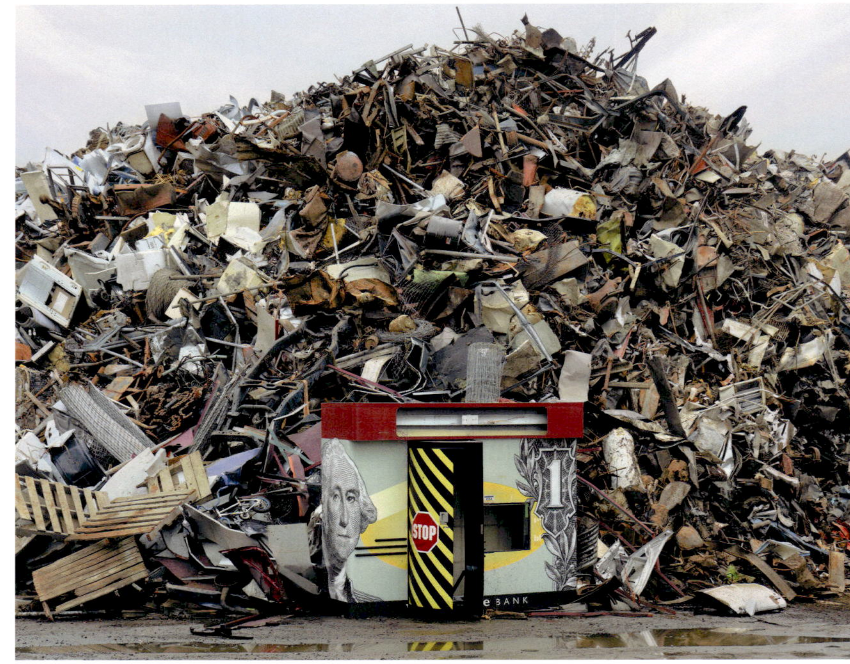

fig. 3: e-Bank, Tacoma, 2004

However, the great metabolic flow of waste that supports our lifestyles is largely unseen. Industry moves, disposes, mines, extracts, shovels, burns, scraps, squanders, and pumps 4 million pounds of material in order to provide one average middle-class American family's needs for a year, more than 1 million pounds of materials per person per year. This includes: 3.5 billion pounds (920 million square yards) of carpet, 2.5 trillion pounds of carbon dioxide gases, 19 billion pounds of polystyrene peanuts, 28 billion pounds of food discarded at home, 360 billion pounds of organic and inorganic chemicals used for manufacturing and processing, 710 billion pounds of hazardous waste generated by chemical production, and 3.7 trillion pounds of construction debris. The figure does not include wastes emitted as the result of extracting gas, coal, oil, and minerals, which add another 16.8 billion tons per year.

Solid wastes in the United States exceed 50 trillion pounds a year. If wastewater is factored in, the annual flow is 250 trillion pounds. Less than 2 percent of the solid waste stream is recycled—primarily paper, glass, plastic, aluminum, and steel. Over the course of a decade, 500 trillion pounds of American resources are transformed into what is revealed by the images in this book. A trillion is a large number: to count to 500 trillion at the rate of one per second would require the entire lifetimes of 240,000 people, the kind of math and sense of scale that Jordan teases out with his twenty-first-century pointillist landscapes. Staring fixedly and gawping at Jordan's work, one cannot help but feel stupefied by the sum shadow of our collective metabolism. This is a kind of activism that demonstrators can only dream about: an instant recognition of the absurdity of how we live.

Jordan's alchemy does more than just shock or cajole. In an ironic way, it mimics nature, for indeed if waste is food in nature, Jordan's depiction of waste is a kind of sustenance, a conscious nourishment that connects our life to the planet, our actions to the collective, our habits to a greater harm. The juxtaposition of the idyllic and the unforgivable, as in Denali the mountain versus the 3-ton, 380-horsepower, 13-mpg Denali

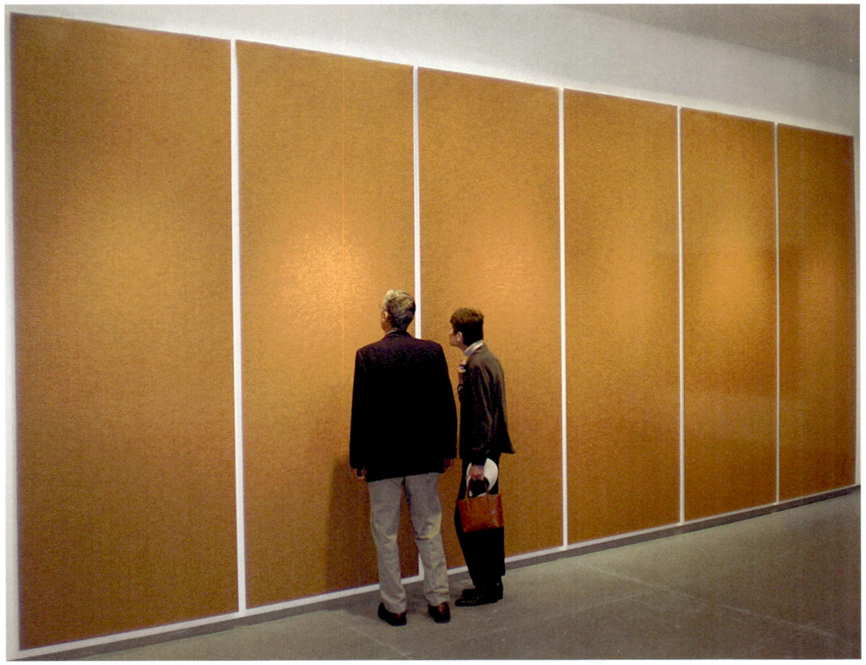

fig. 4: Prison Uniforms, 2007, installed at Von Lintel Gallery, New York, 2007

SUV, is a constant in his work (plate 5). The intrinsic beauty of the natural world is fractured by the quenchless desires of our collective hungry ghost. In every instance, it is our desire we see staring back at us from the images—transformations of the profane that reflect back our unconscious cravings. Thirteenth-century Buddhist writings prophesied such a time and called it the Age of Dregs. There is no other work that captures the catabolic collapse of our industrial destiny as evocatively and clearly.

I find Jordan's work revelatory. I do not want to turn away from it. Though it may depict waste, it is not repugnant. It fascinates. Jordan takes data—"running the numbers," as he puts it—and explodes them into our imagination. The effect is indelible and defiantly beautiful, elegant in a way that was formerly unimaginable. There is not a trace of righteousness or "gotchas" here. It is as if a naturalist took us into the woods to see a nest of hummingbirds or monarch butterfly cocoons finely woven into the canopy, but in this case we are taken into the gross, not the sublime: the graveyards of industry, the mountains of dust and tailings and rust. We become witnesses to our undoing, the unraveling of life. Moreover, this work exposes a form of waste we mostly dissociate ourselves from—the wasting of people, lives, and cultures. Here we see the 2.3 million prison uniforms (plate 13; fig. 4), the tortured inmates of Abu Ghraib, the uninsured children who do not get proper healthcare, an unending war, handguns, the infinitude of tobacco. This connection between wasting our planet and wasting each other is the denouement. How we treat each other is how we treat the earth. One hundred million people wander the planet because they are refugees, displaced, or homeless, while 650 million cars rove the earth double glazing the upper stratosphere with carbon emissions.

What Jordan is doing is a benediction, a gift to the world. It is a kind of sacred art, not because it unveils the hidden world of spirit and devotion but because it brings us back to our senses. It is the Book of Dross, wordlessly posing timeless questions about who we want to be in this fragile and brief lifetime. Buckminster Fuller observed that spaceship earth is so ingeniously designed that we have no idea we are on one. Call it what you wish, any metaphor will do as long as it honors the simple fact that we are on an orb hurtling along at a million miles an hour and nothing can be thrown away without diminishing life itself: not our future, not our resources, not our social covenant. Our hearts will break over and again if we destroy who we are, whether it is our children or our homeland. Jordan's images are a prayer, a supplication to all who look upon them, to harm no more, to be mindful in all that we do, speak, and take.

Paul Hawken is an environmentalist, entrepreneur, journalist, and best-selling author. His recent book is Blessed Unrest: How the Largest Movement in the World Came into Being and Why No One Saw It Coming.

RUNNING THE NUMBERS
AN AMERICAN SELF-PORTRAIT

This series looks at contemporary American culture through the austere lens of statistics. Each image portrays a specific quantity of something: thirty seconds of our aluminum can consumption, eight hours of commercial jet flights, and so on. My hope is that images representing these statistics might have a different effect than the raw numbers we encounter daily in books, magazines, and the news. Statistics about our mass culture can feel abstract and anesthetizing, making it hard to connect with the profoundly important issues they represent.

Finding meaning in these huge numbers is made even more difficult by the fact that there is nowhere we can go to see the actual phenomena—the millions of barrels of oil we burn daily, or the eleven hundred Americans who die every day from smoking cigarettes, or all the children in our country who lack health insurance. They are spread across our nation in an invisible collective that is impossible to experience in any way other than by rows of numbers with lots of zeros, written on a page.

In this way, I think of *Running the Numbers* as a kind of translation, from the deadening language of statistics into a more universal visual language that might allow for more feeling. The underlying aim is to question our roles and responsibilities as individuals in a society that is increasingly enormous, incomprehensible, and overwhelming.

~cj (2008)

Plastic Cups, 2008 (detail of plate 16)

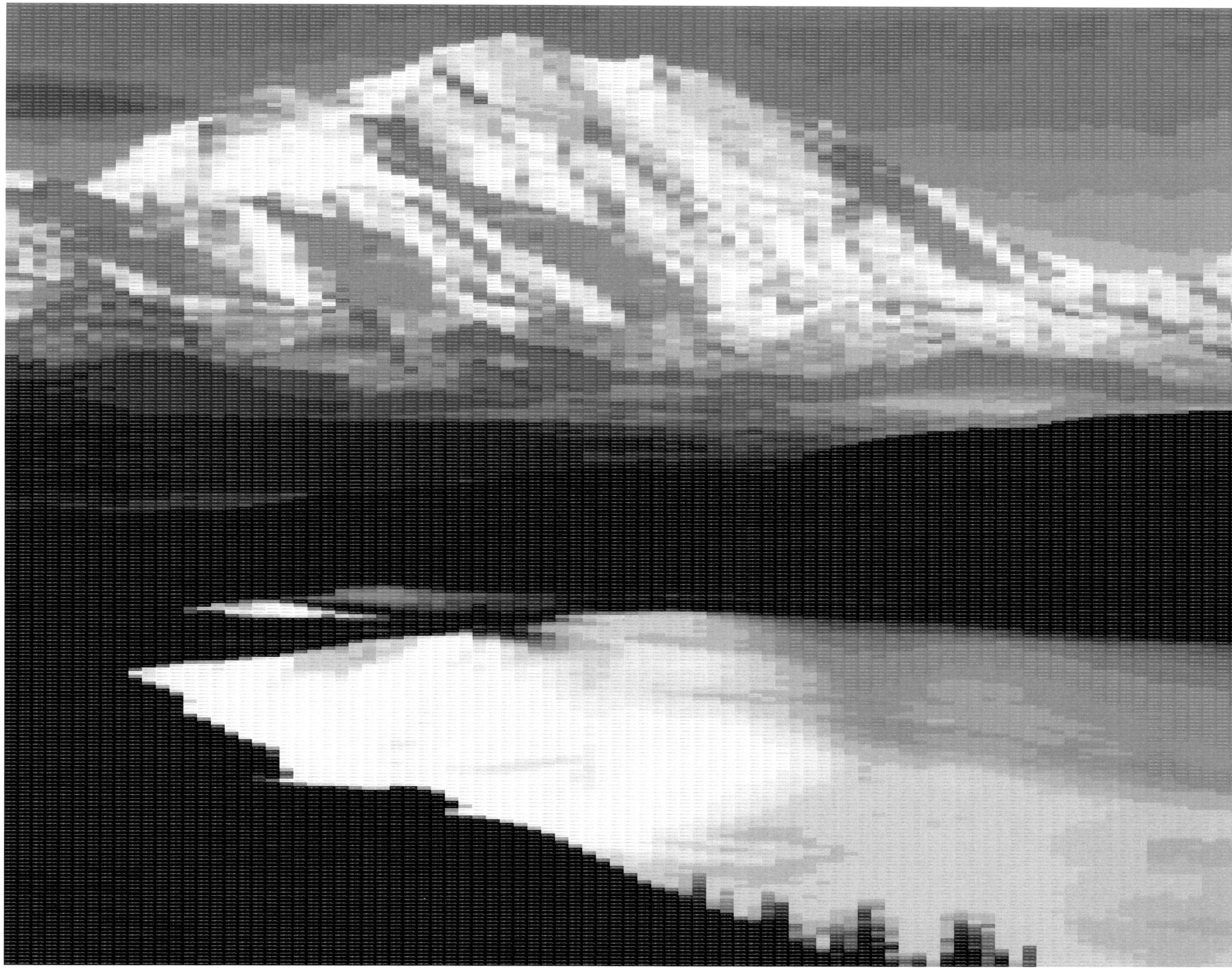

Depicts 24,000 logos from the GMC Yukon Denali, equal to six weeks of sales of that model SUV in 2004

(plate 5) Denali Denial, 2006

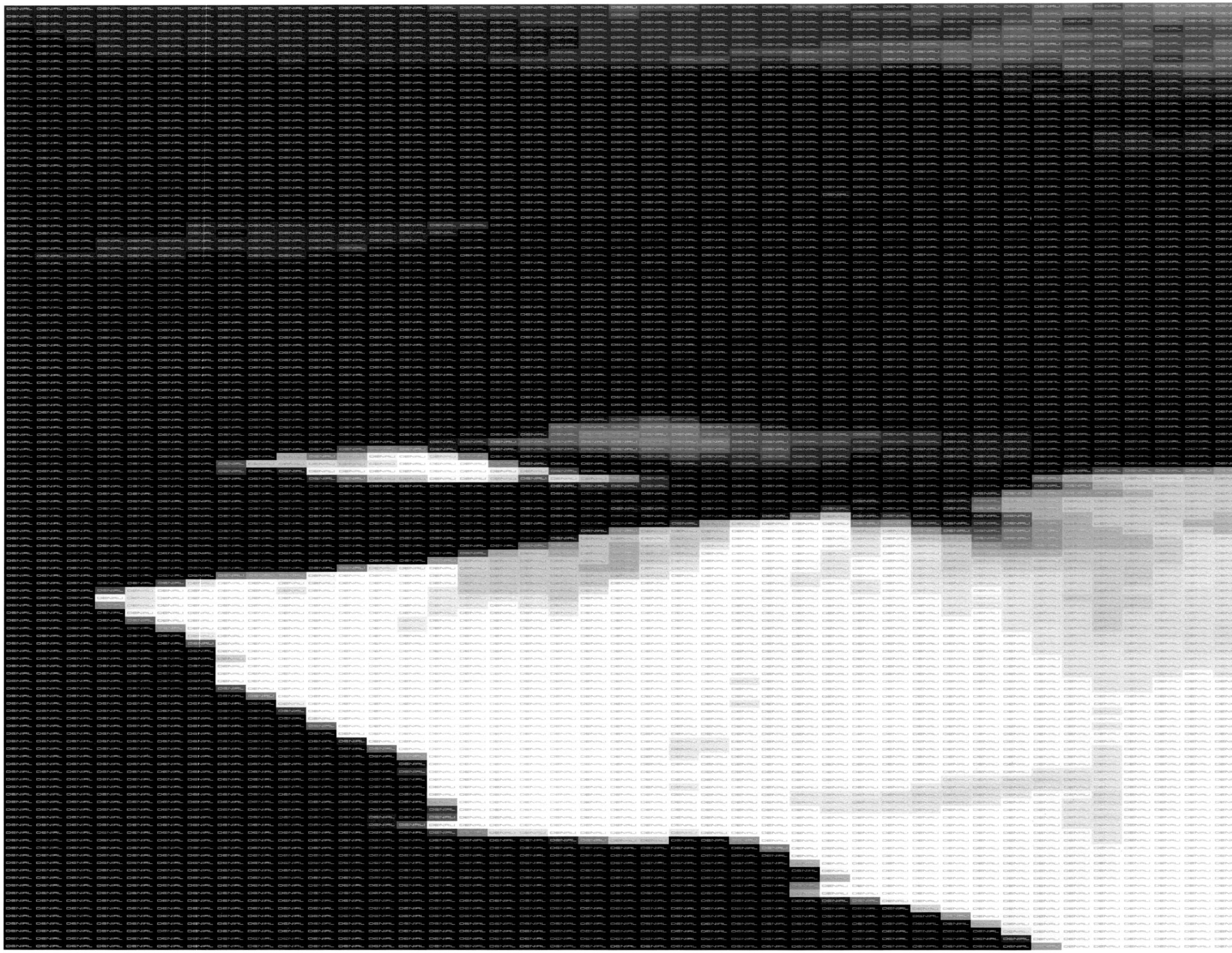
DENIAL

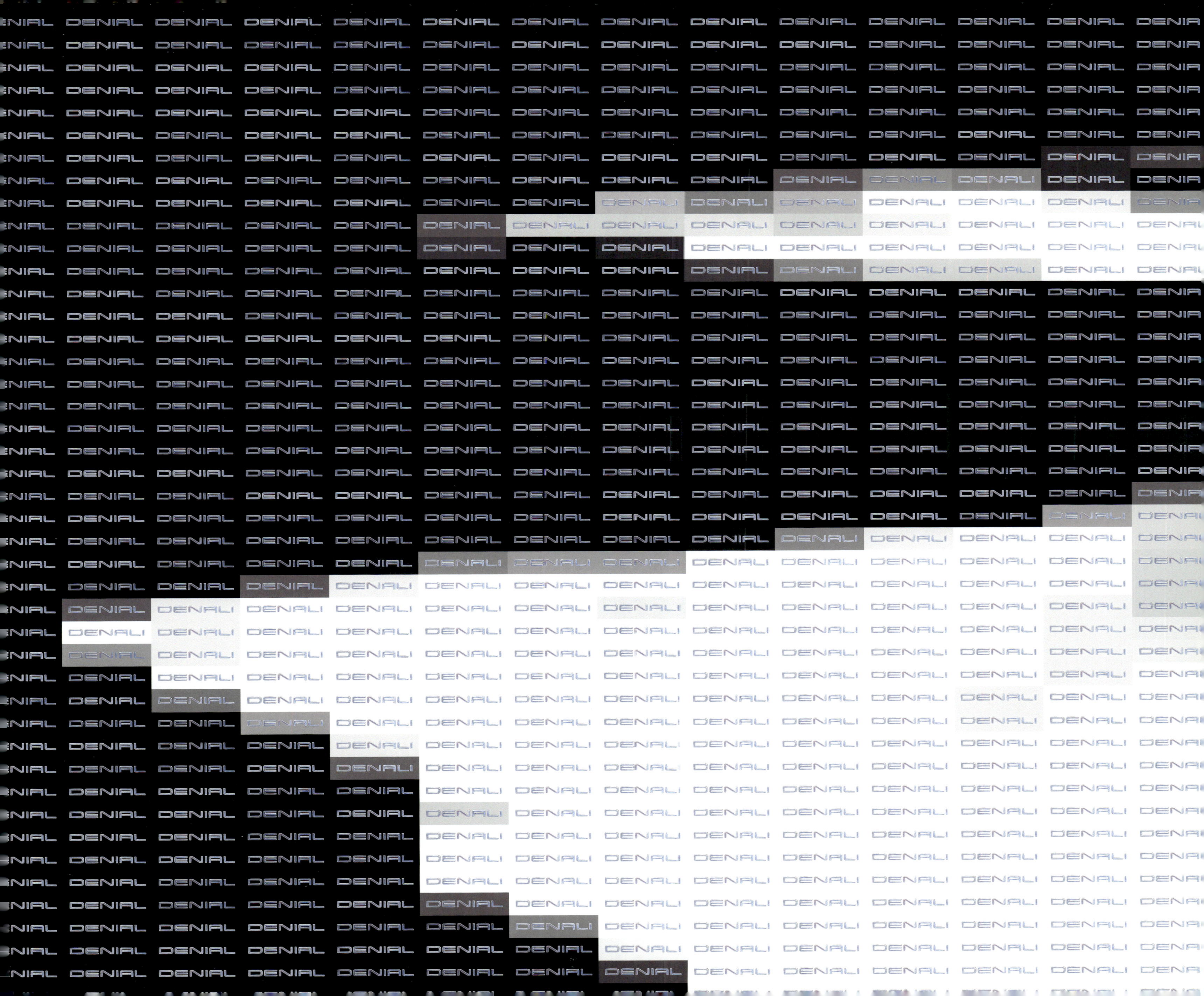
DENIAL
DENALI

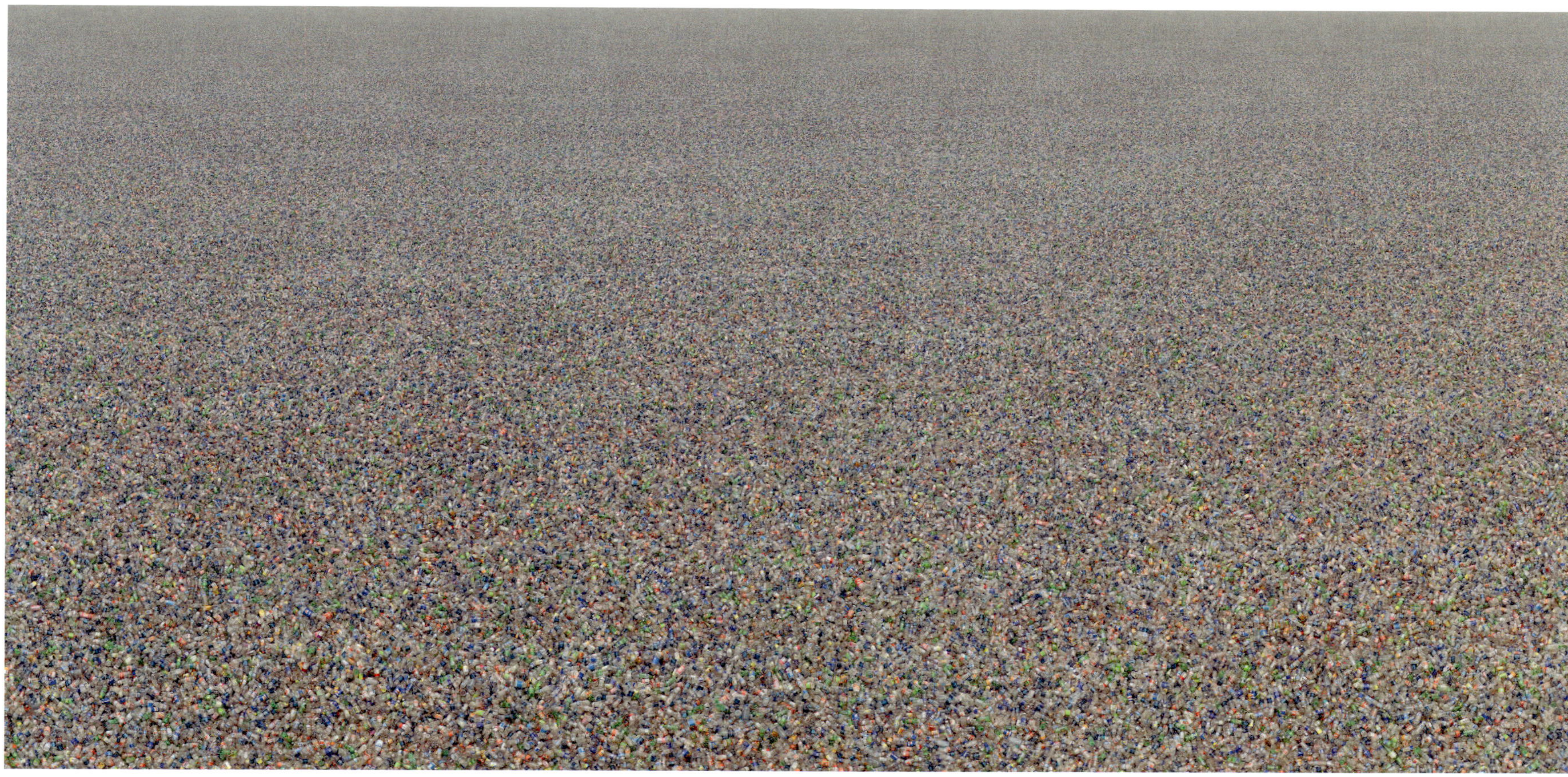

Depicts two million plastic beverage bottles, the number used in the United States every five minutes

(plate 6) Plastic Bottles, 2007

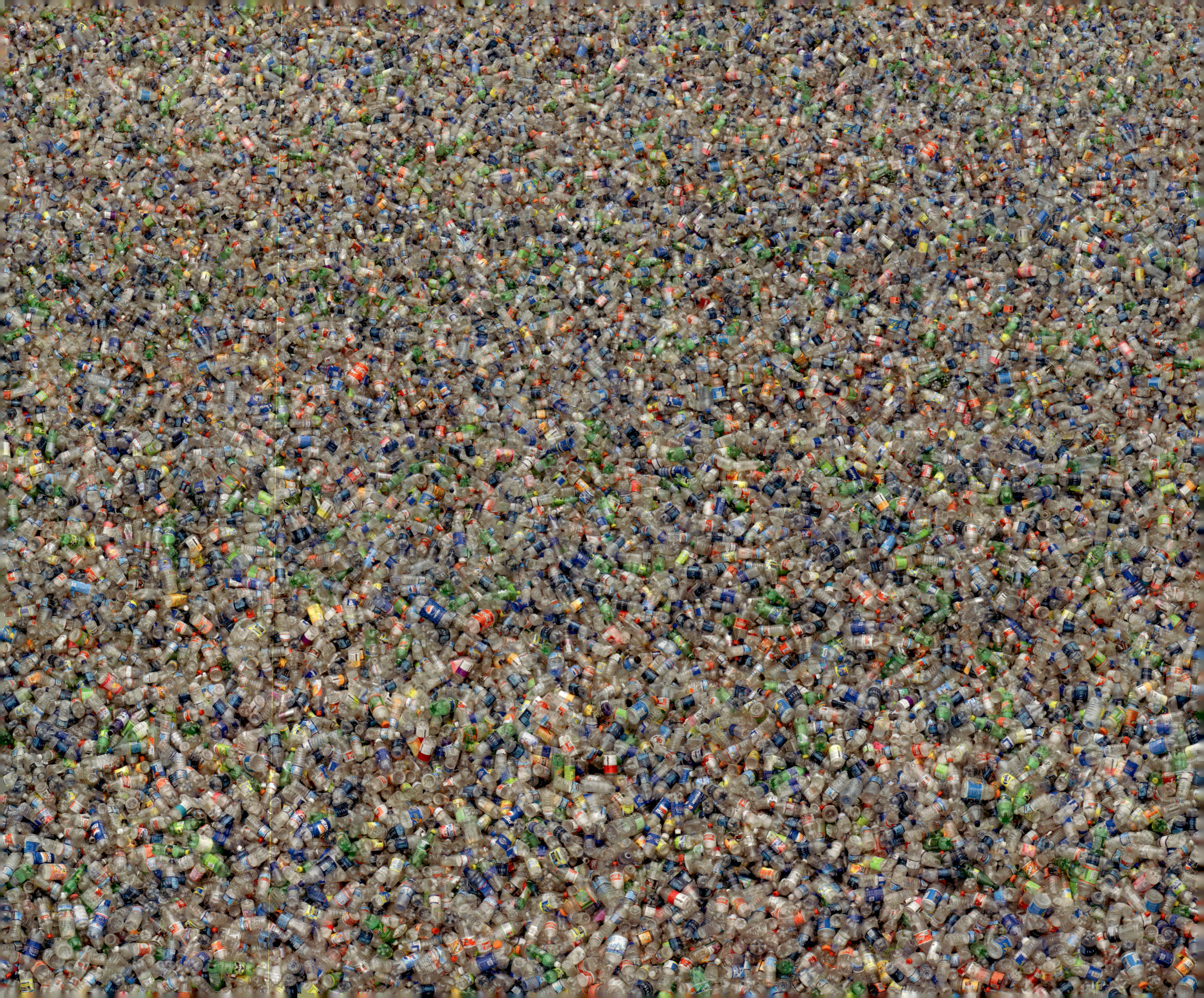

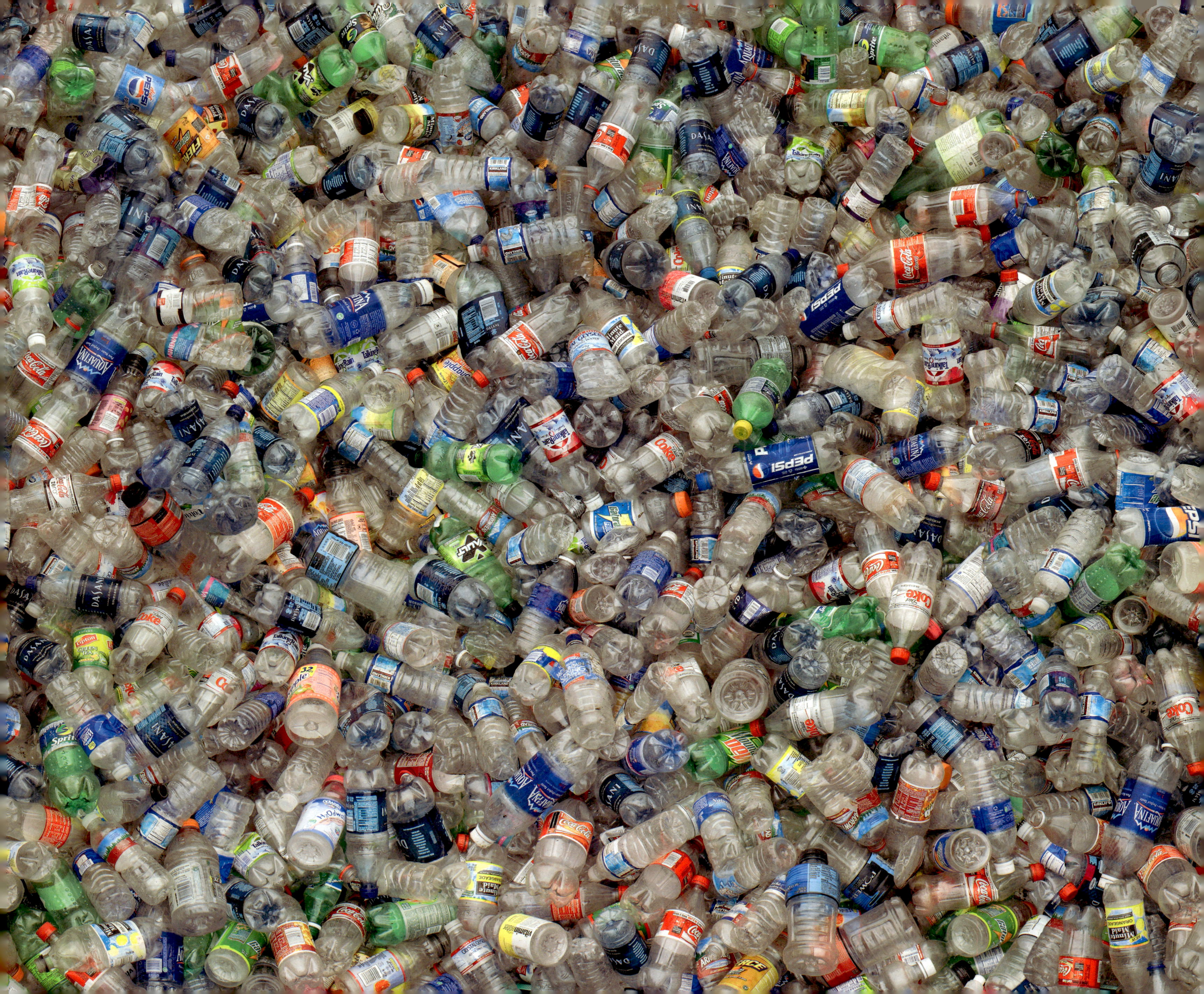
PEPSI

Depicts 426,000 cell phones, the number retired in the United States every day

(plate 7) Cell Phones, 2007

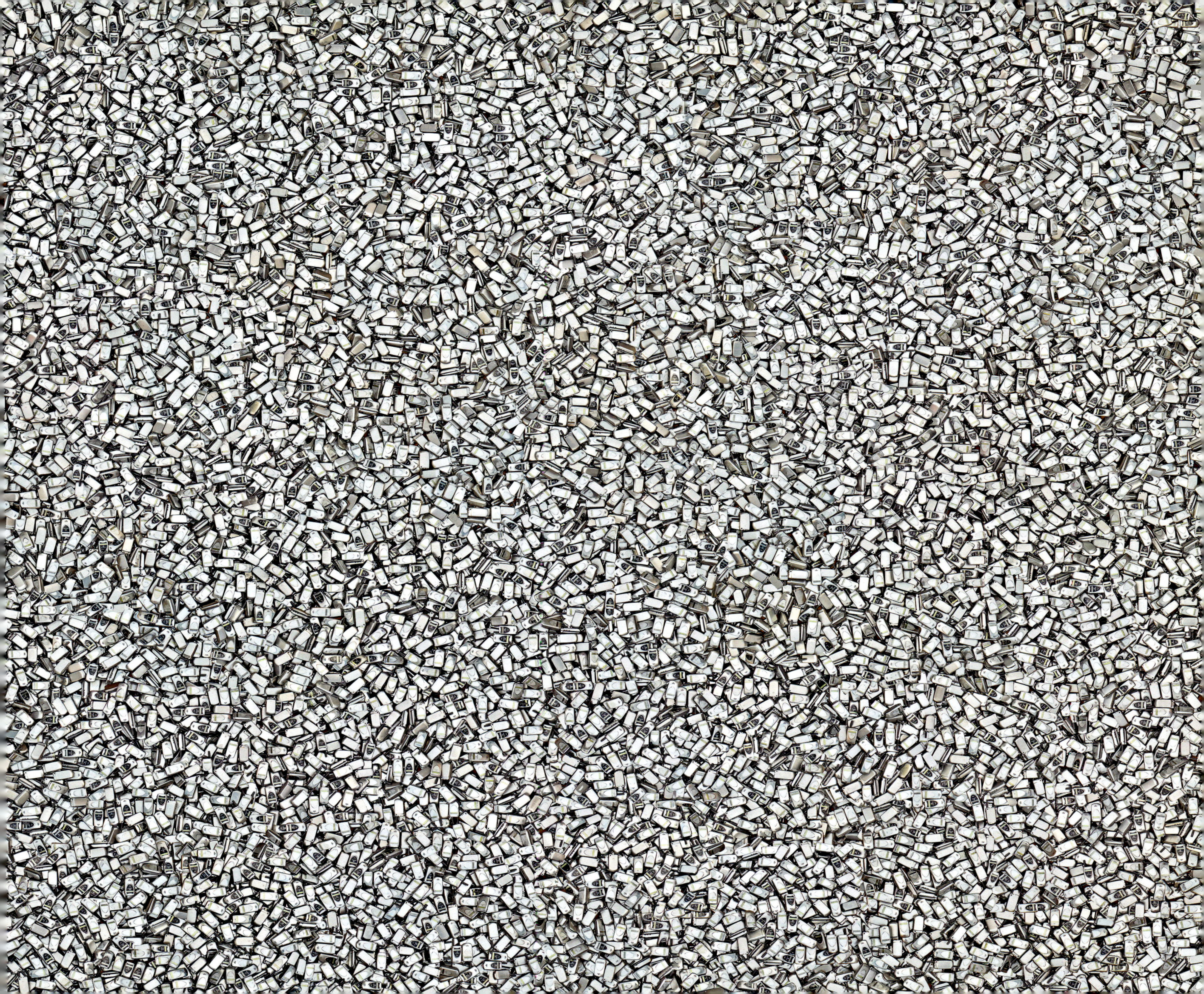

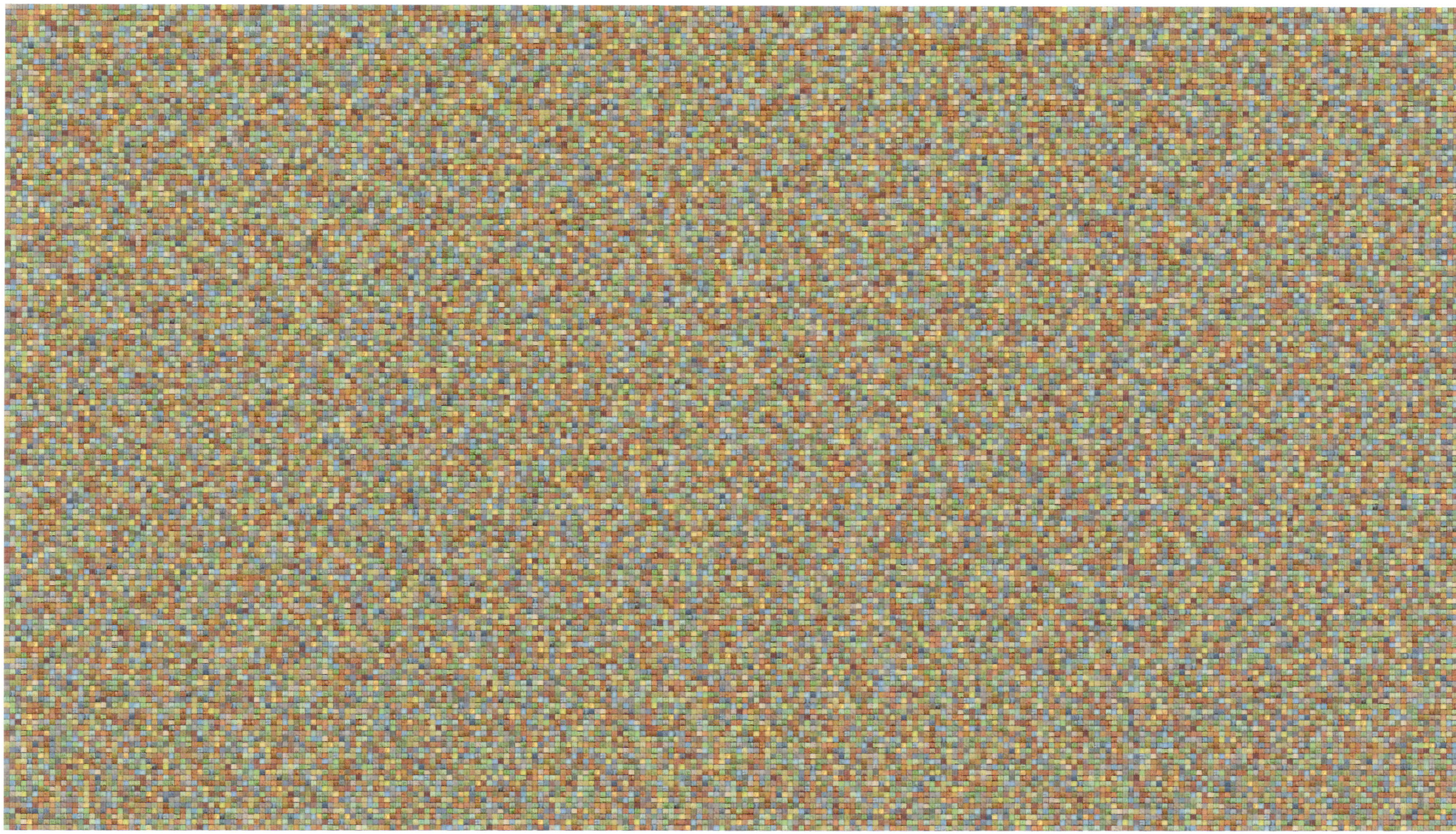

Depicts 38,000 shipping containers, the number processed through American ports every twelve hours

(plate 8) Shipping Containers, 2007

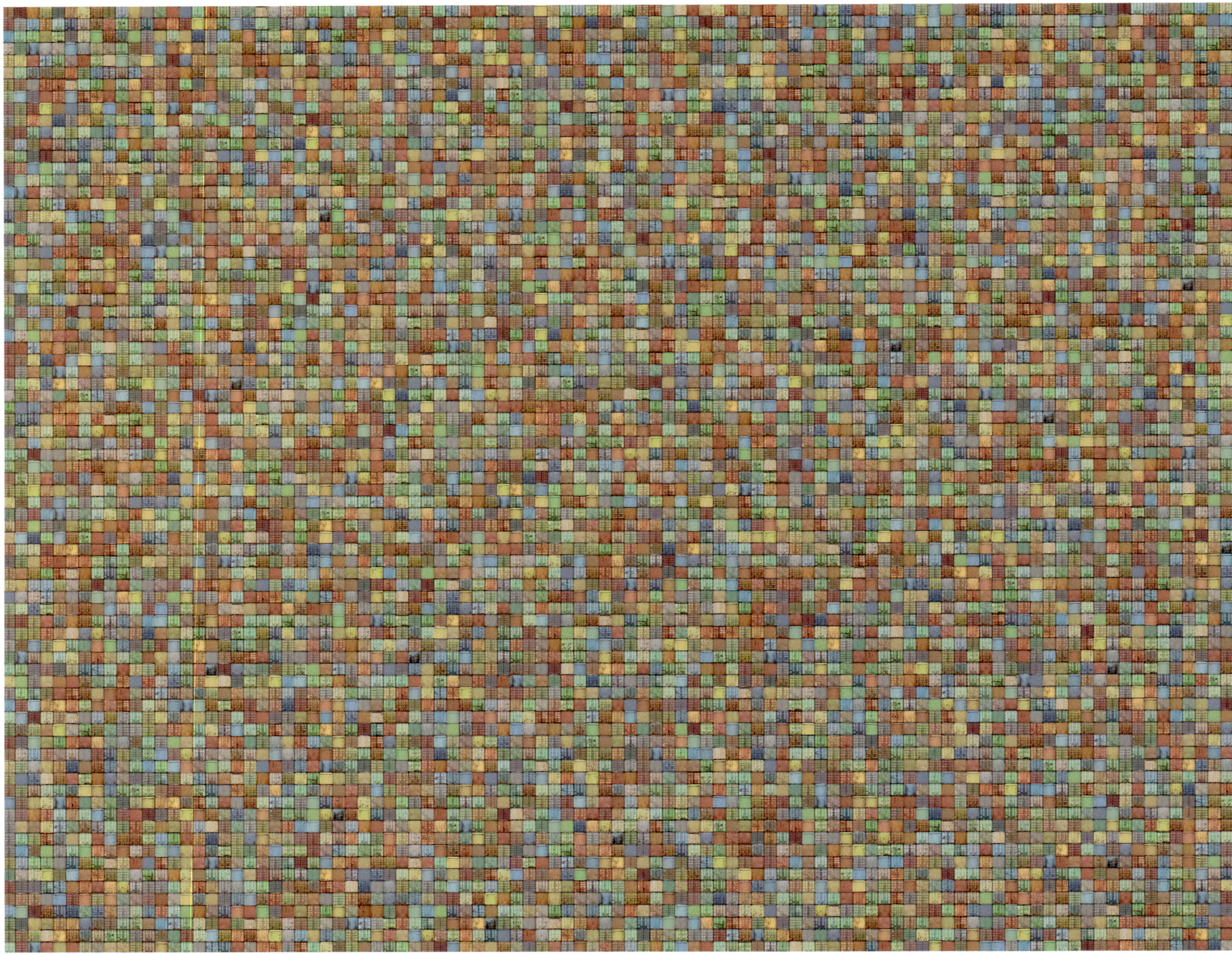

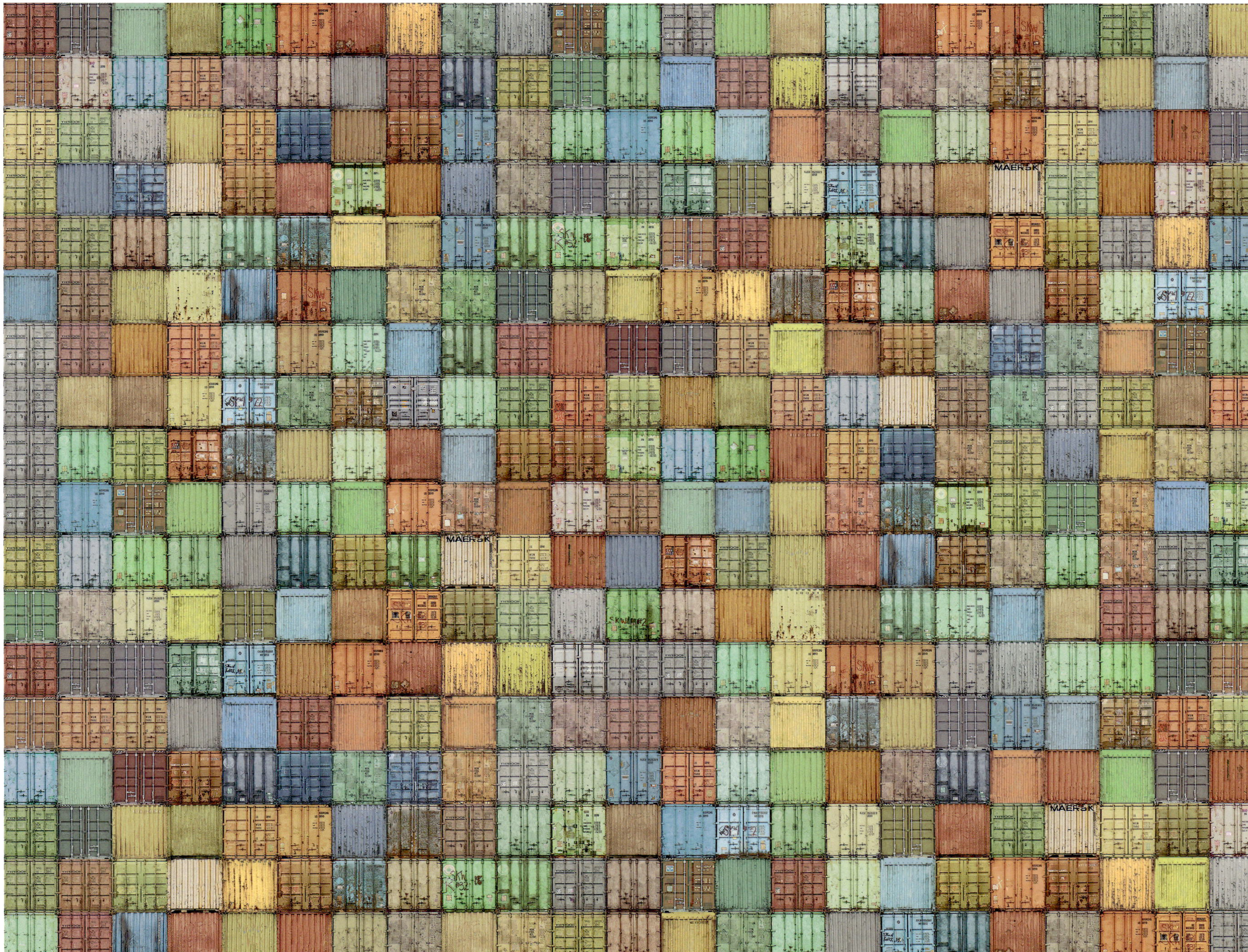
MAERSK
MAERSK
MAERSK

Depicts 29,569 handguns, equal to the number of gun-related deaths in the United States in 2004

(plate 9) Handguns, 2007

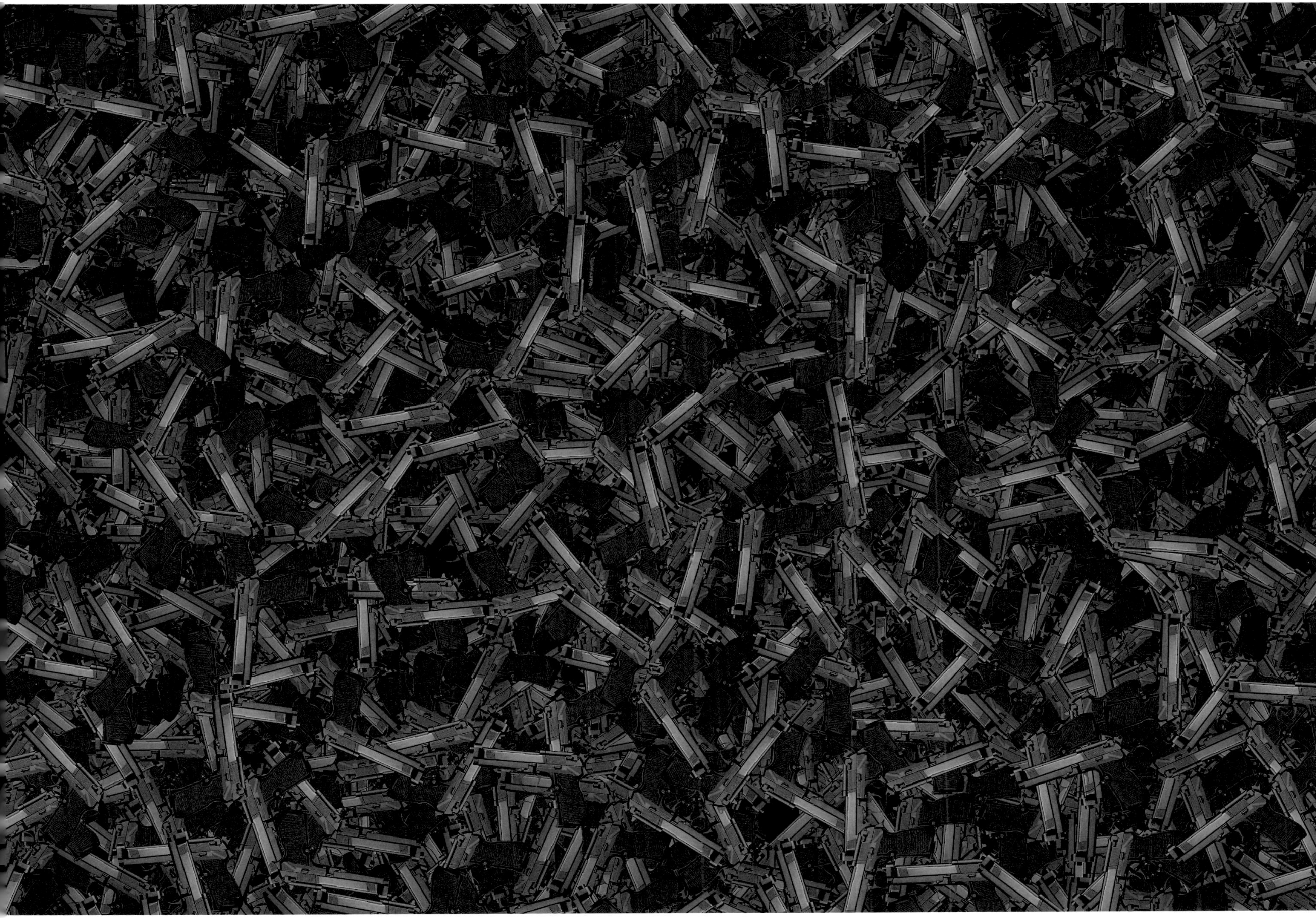

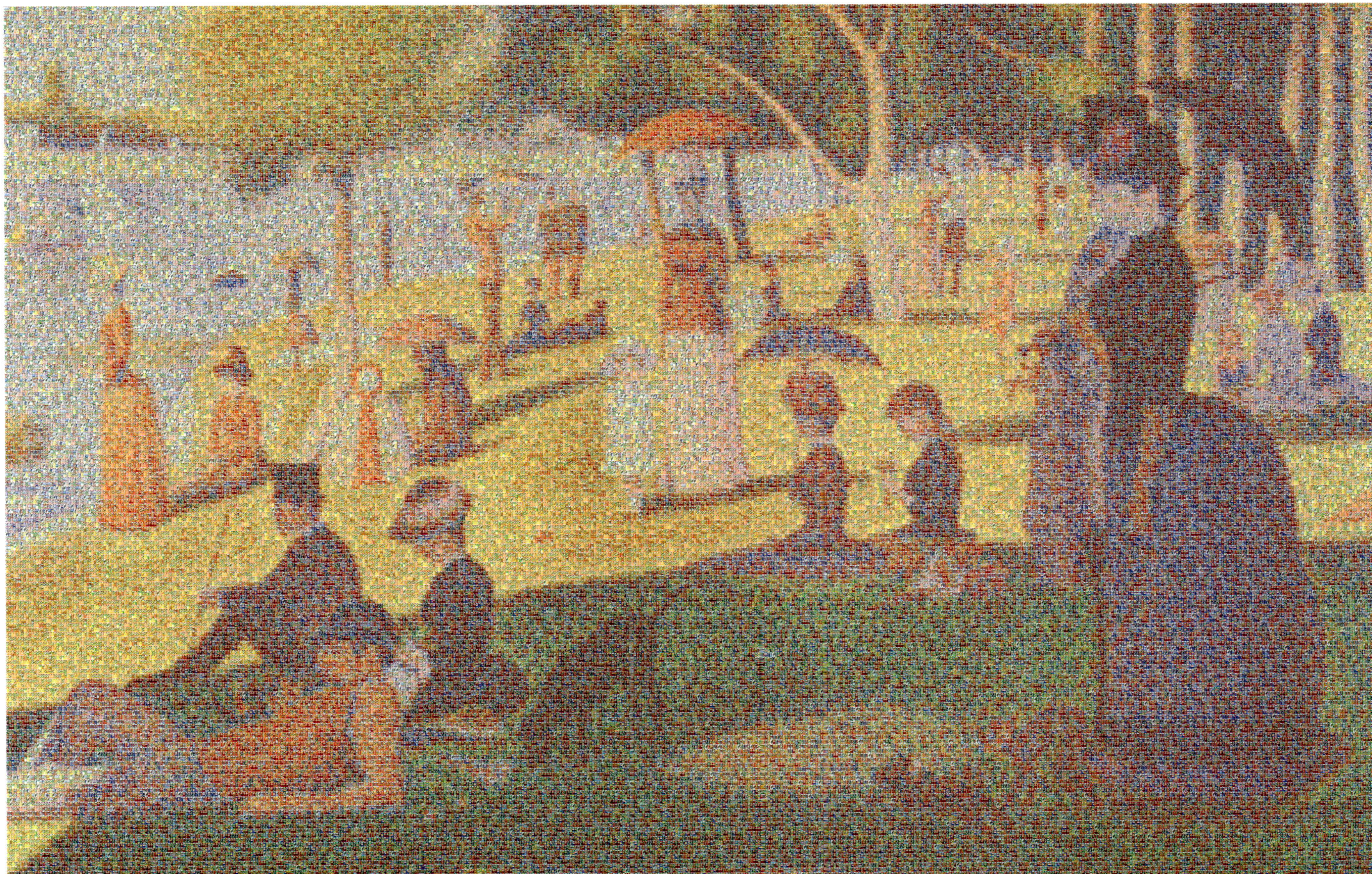

Depicts 106,000 aluminum cans, the number used in the United States every thirty seconds

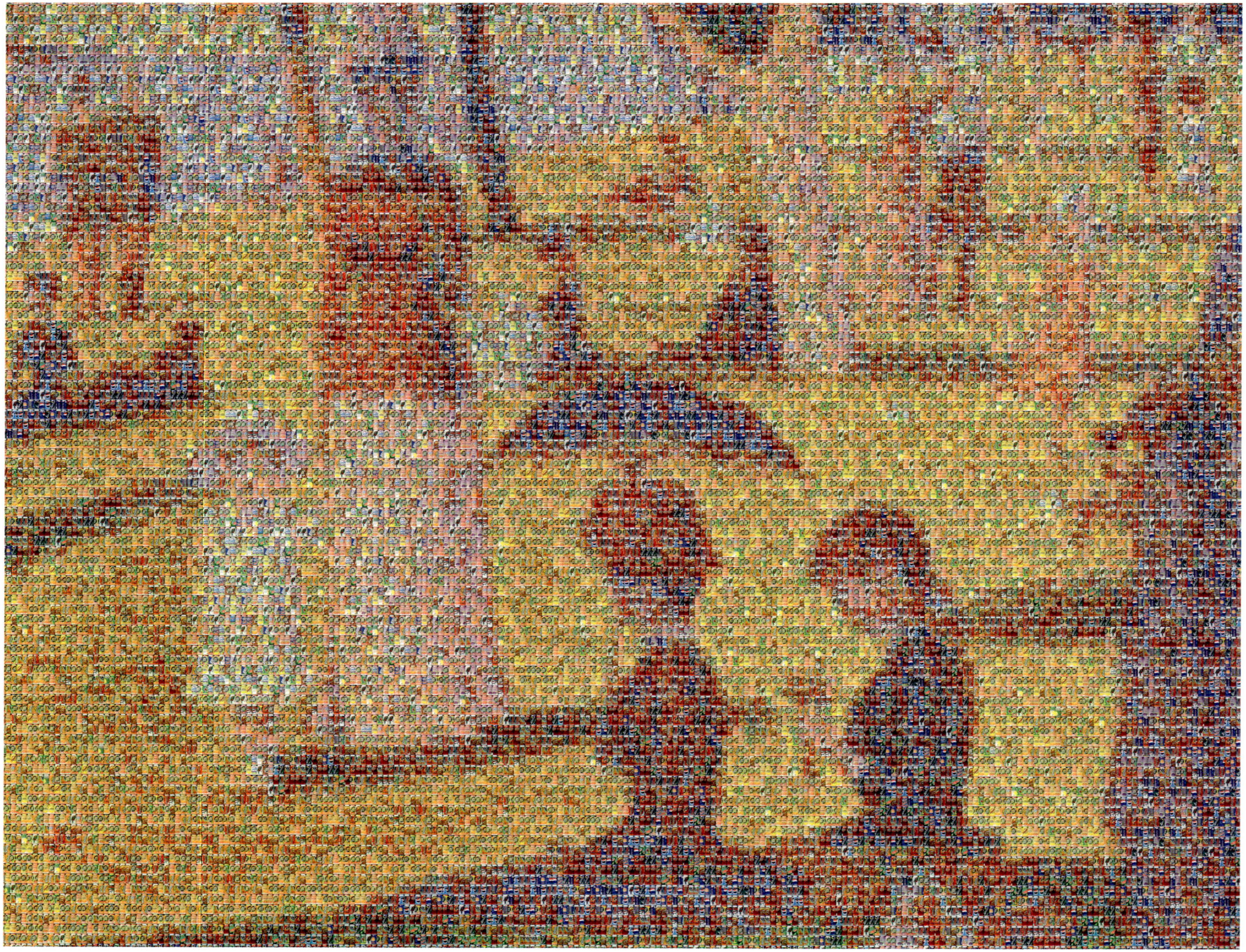

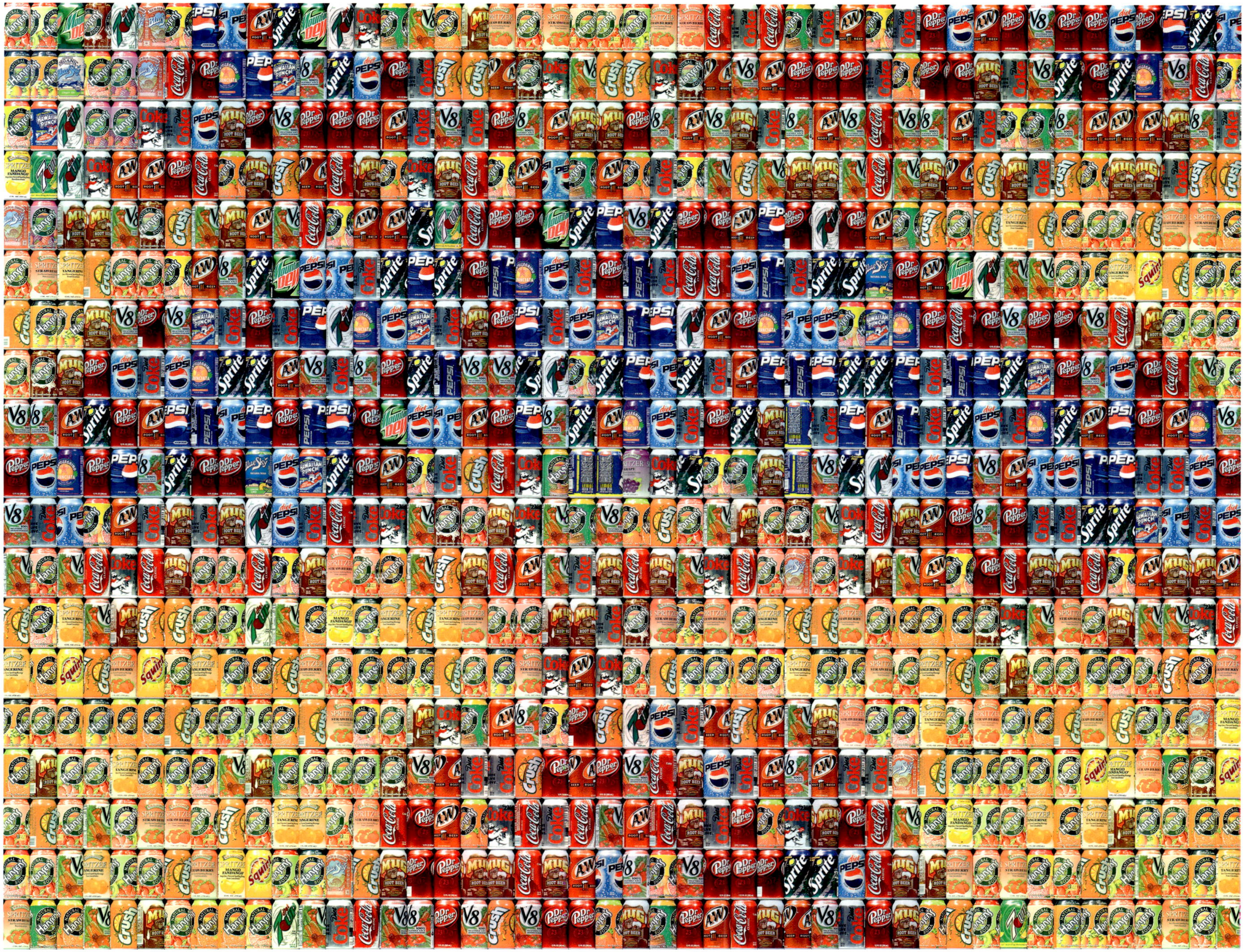

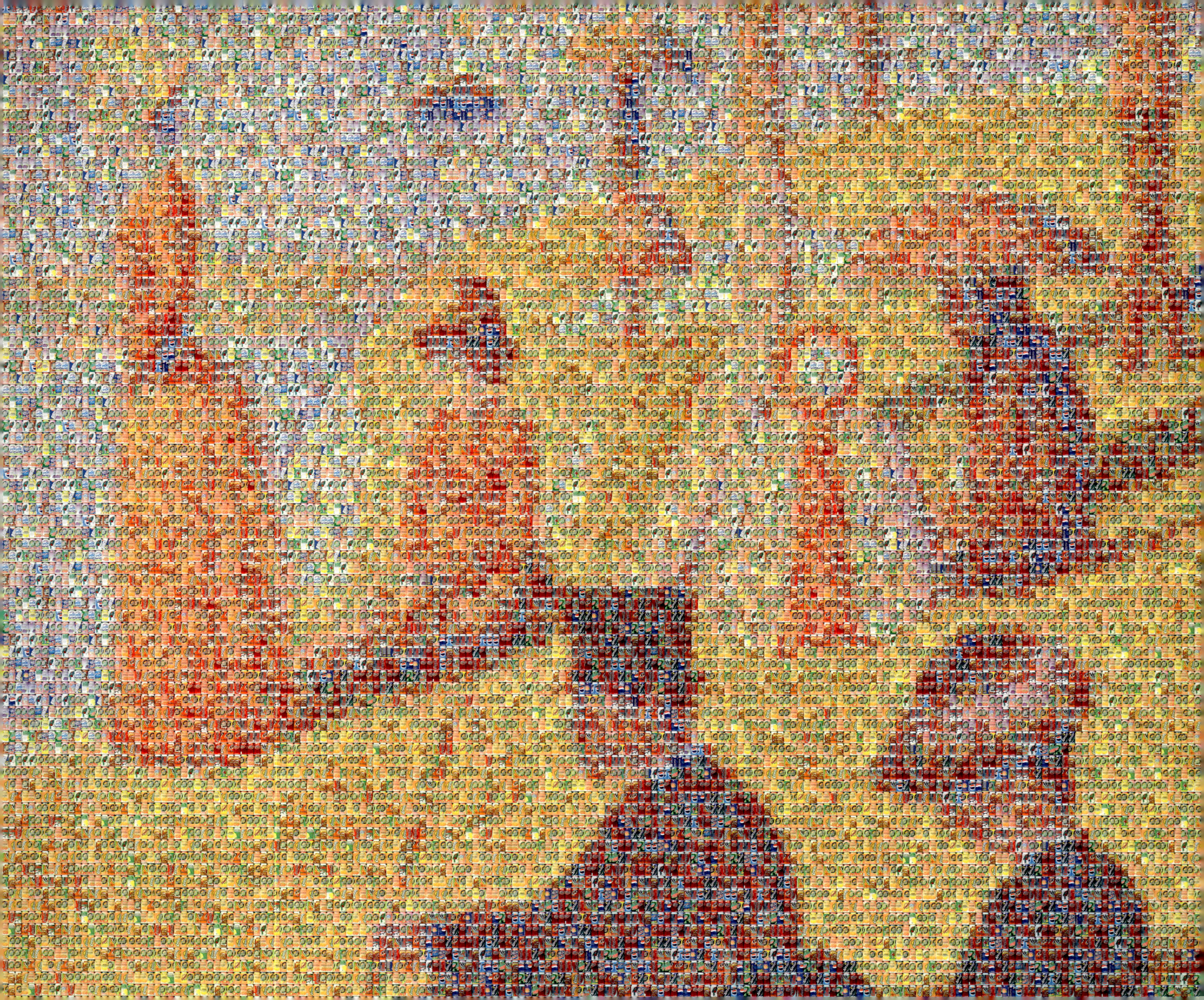

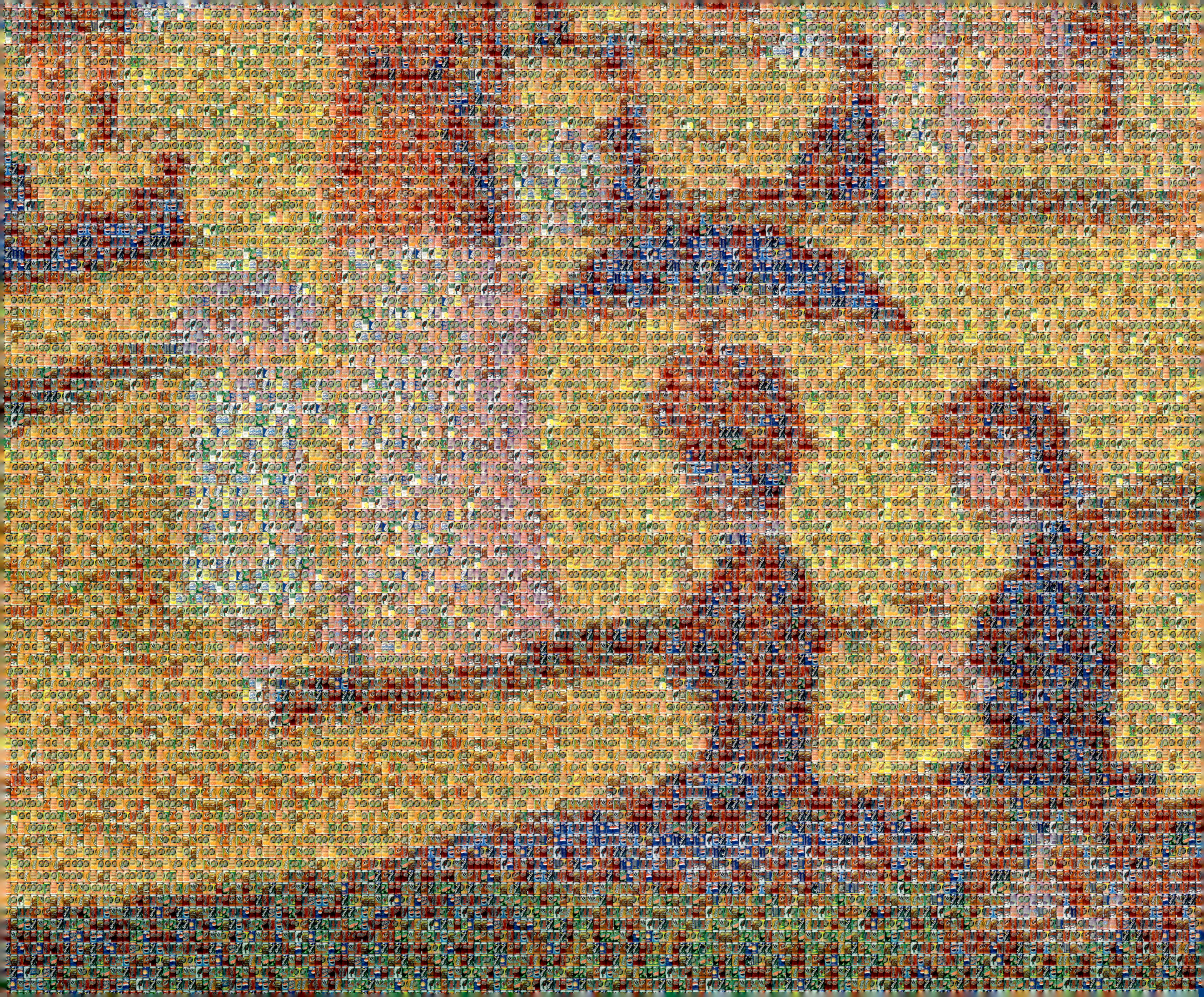

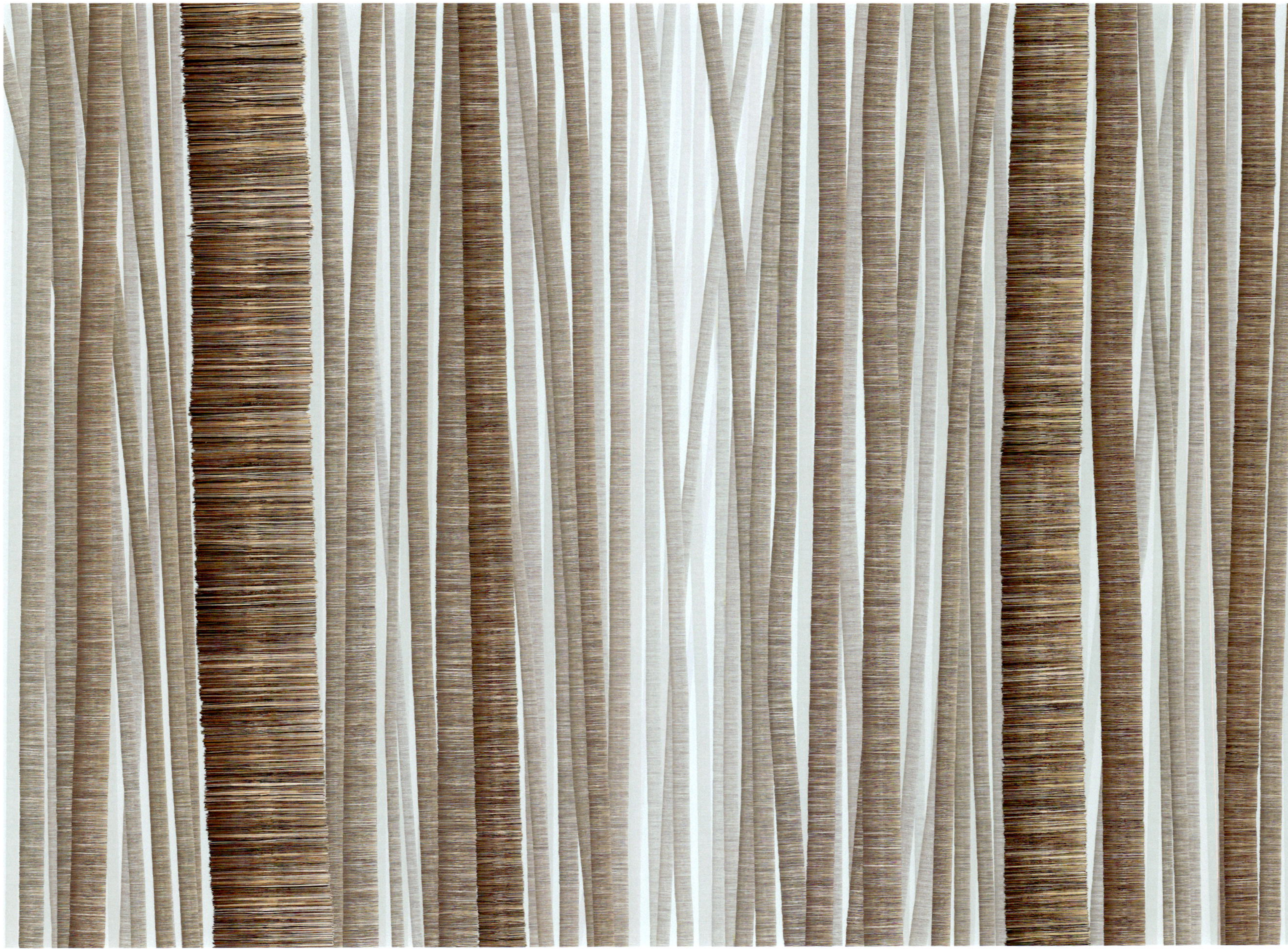

Depicts 1.14 million brown paper supermarket bags, the number used in the United States every hour

(plate 11) Paper Bags, 2007

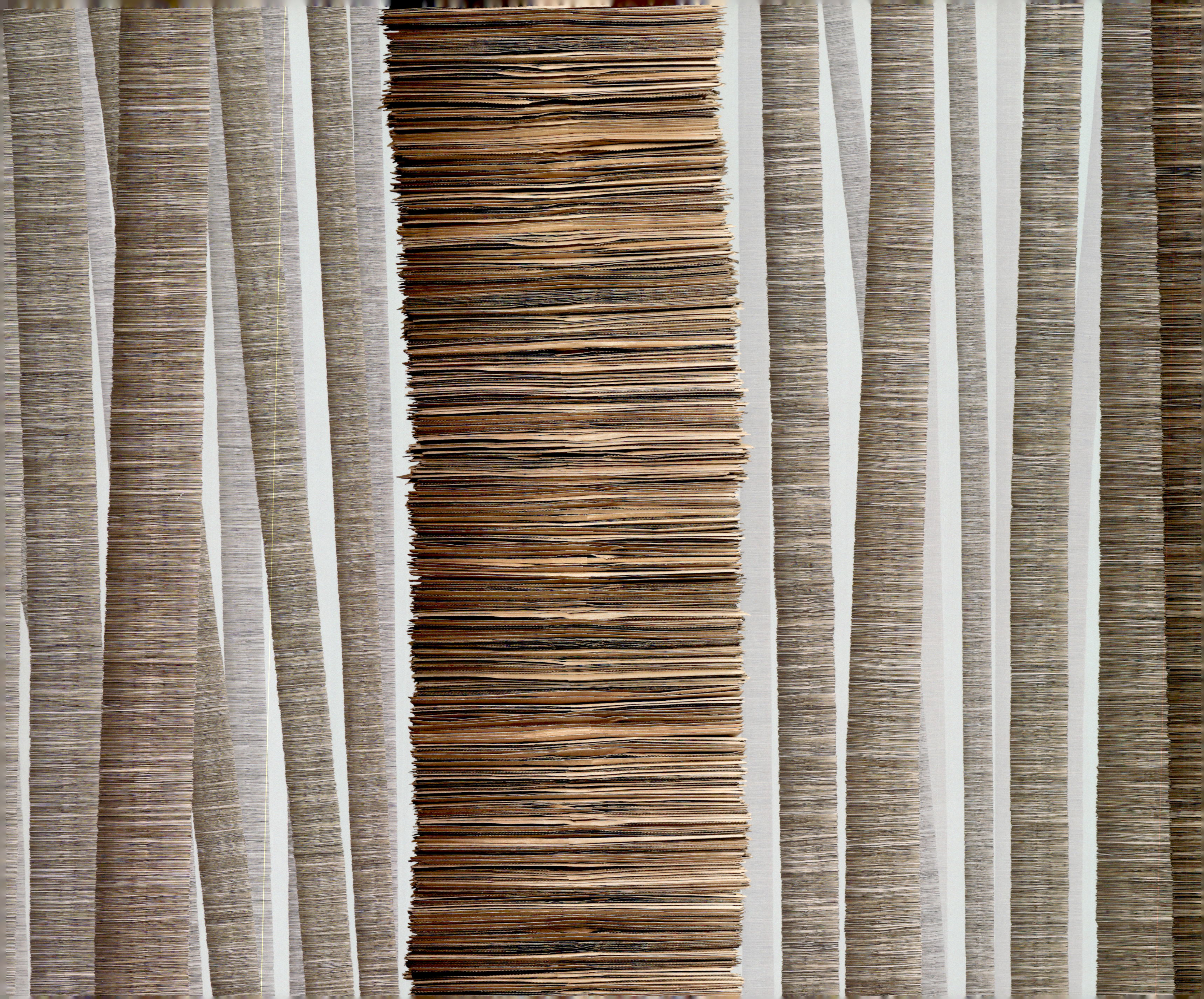

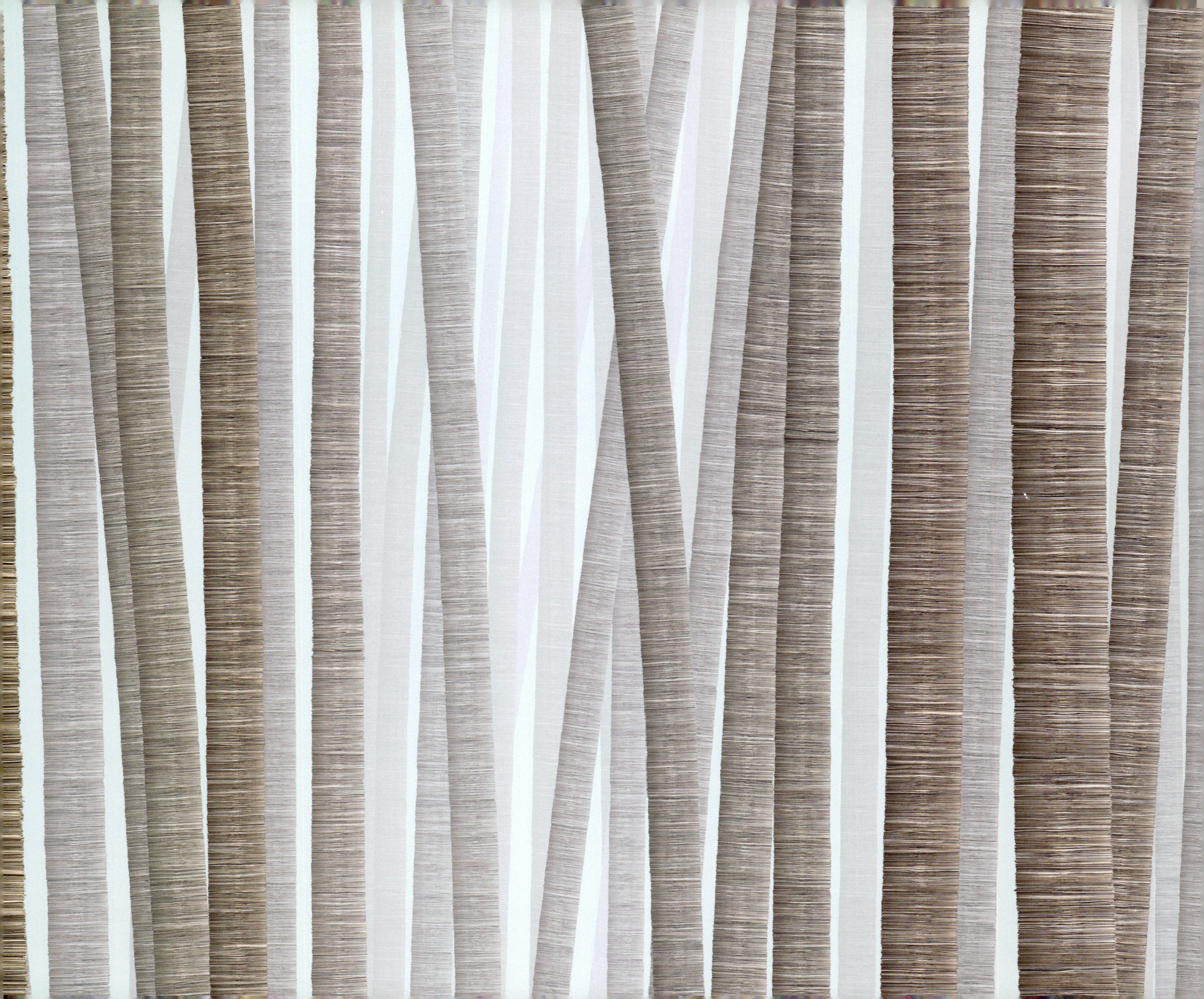

Depicts 11,000 jet trails, equal to the number of commercial jet flights in the United States every eight hours

(plate 12) Jet Trails, 2007

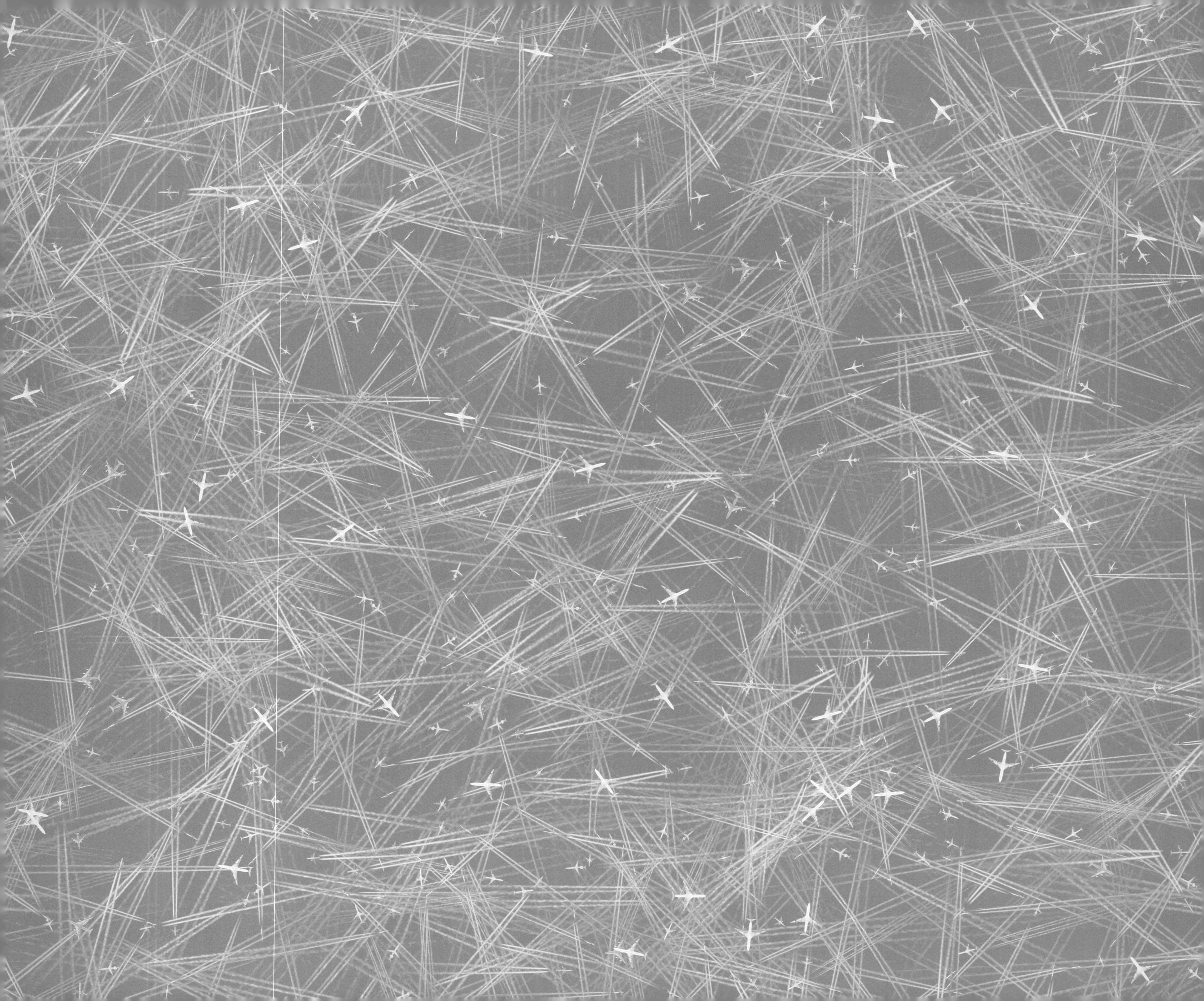

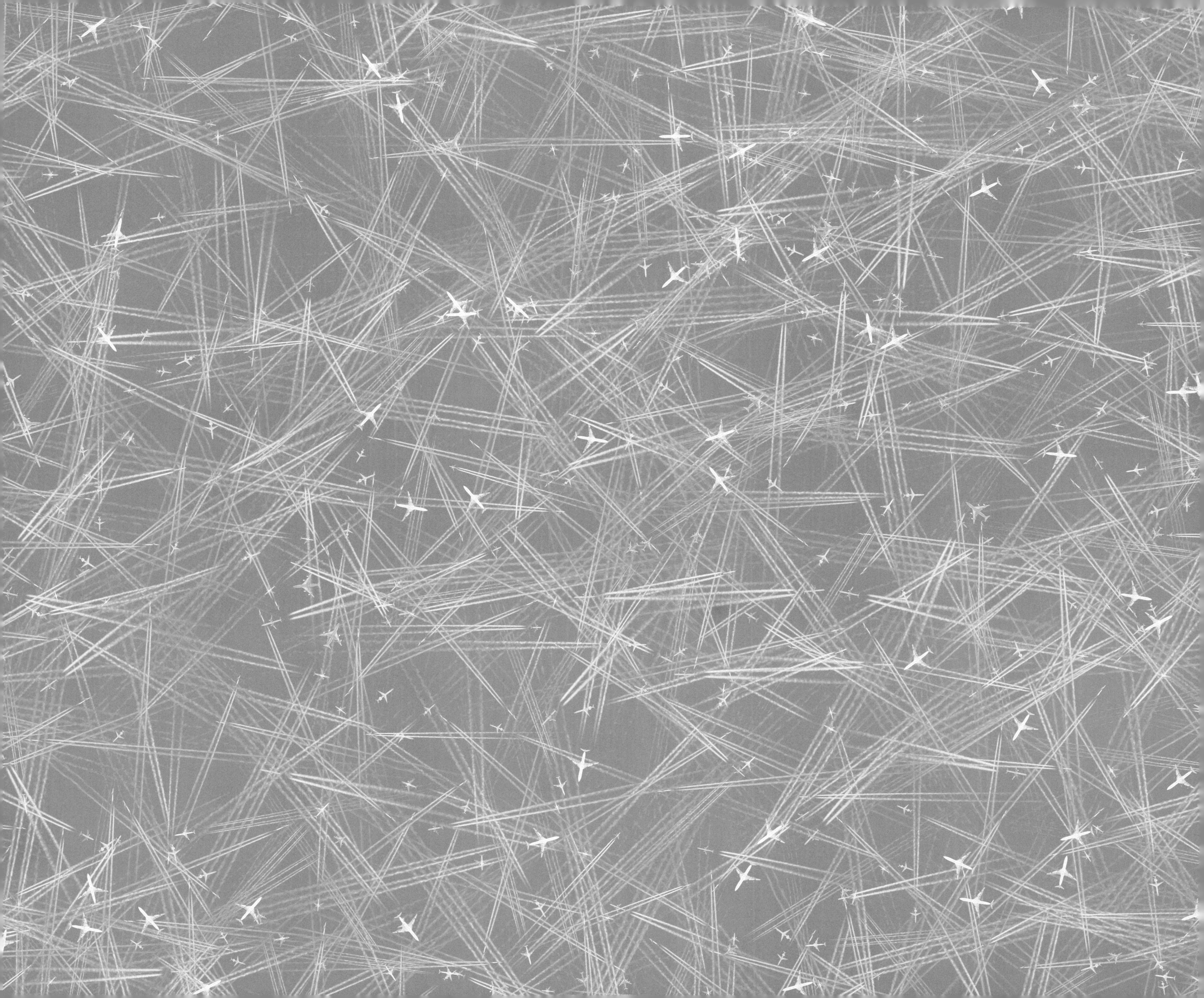

Depicts 2.3 million folded prison uniforms, equal to the number of Americans incarcerated in 2005

(plate 13) Prison Uniforms, 2007

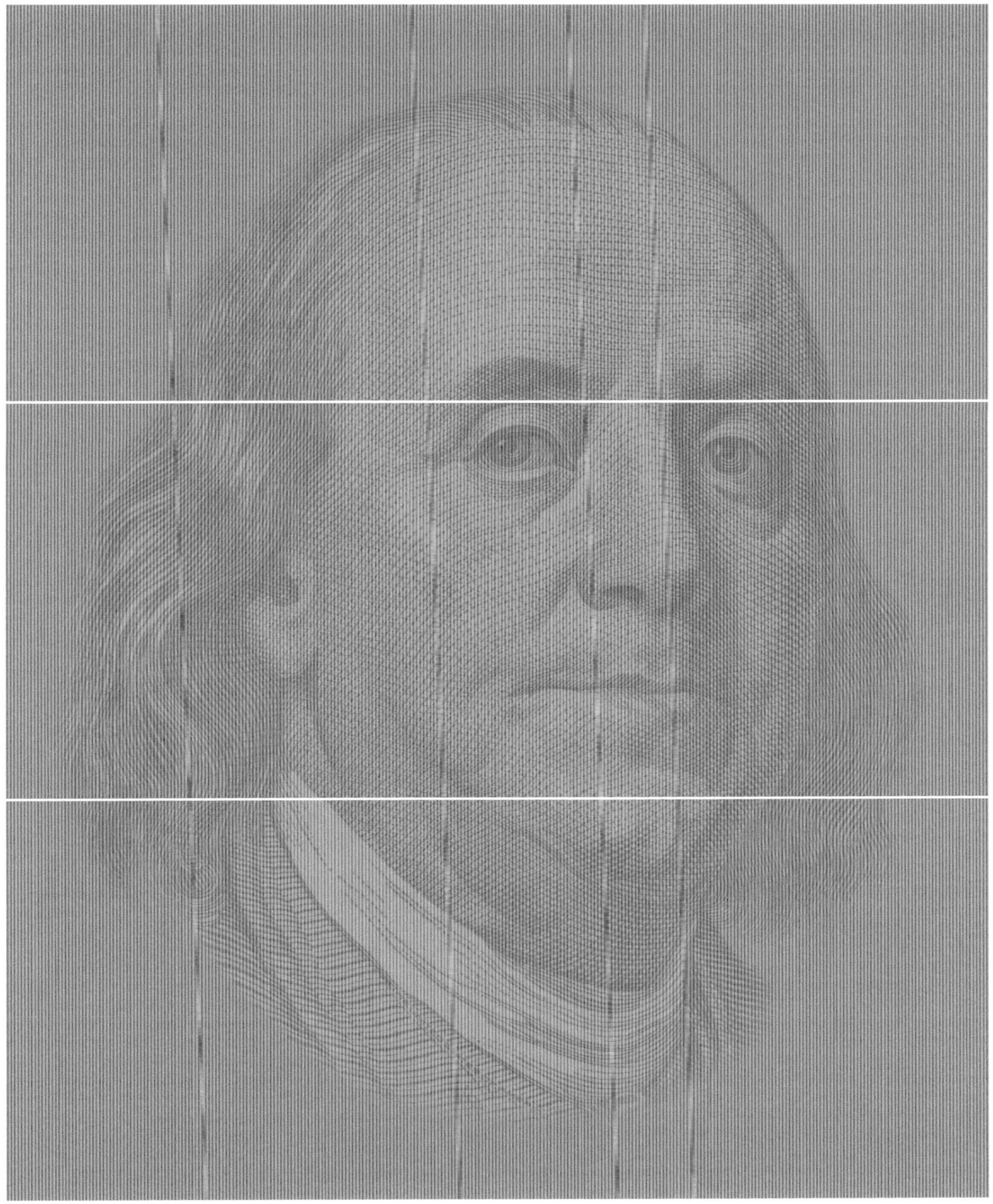

Depicts 125,000 $100 bills ($12.5 million), the amount our government spent every hour on the war in Iraq from 2003 through 2008

(plate 14) Ben Franklin, 2007

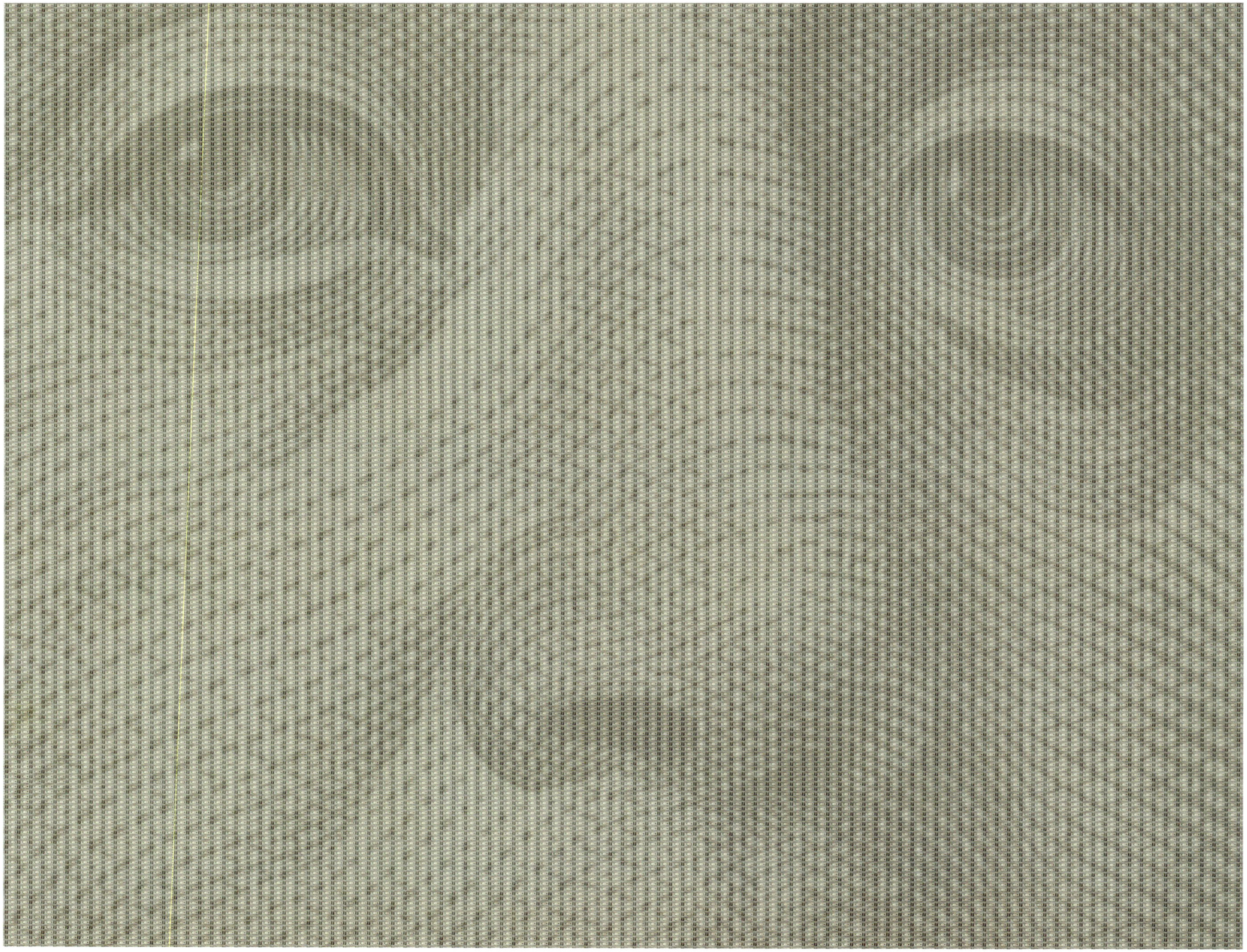

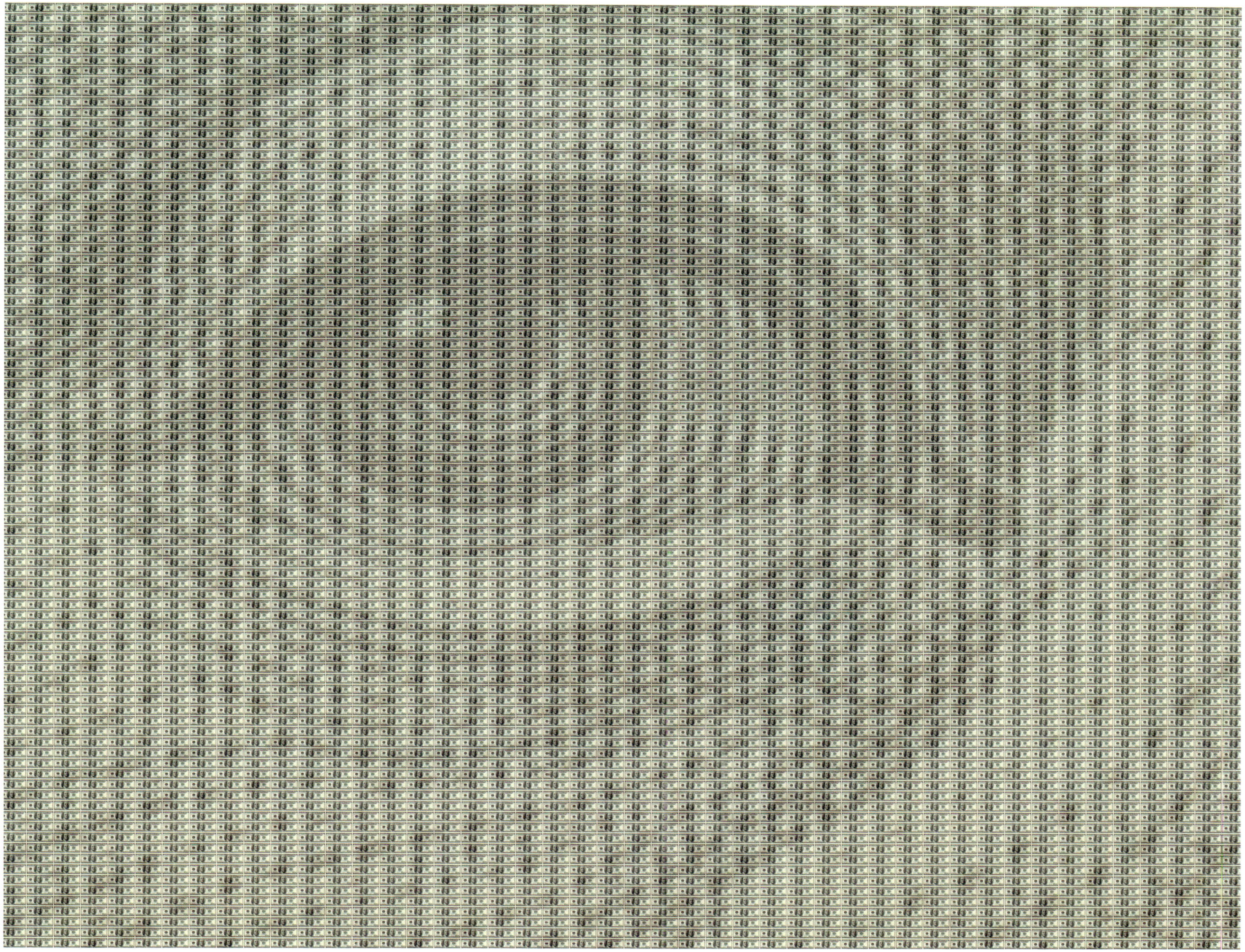

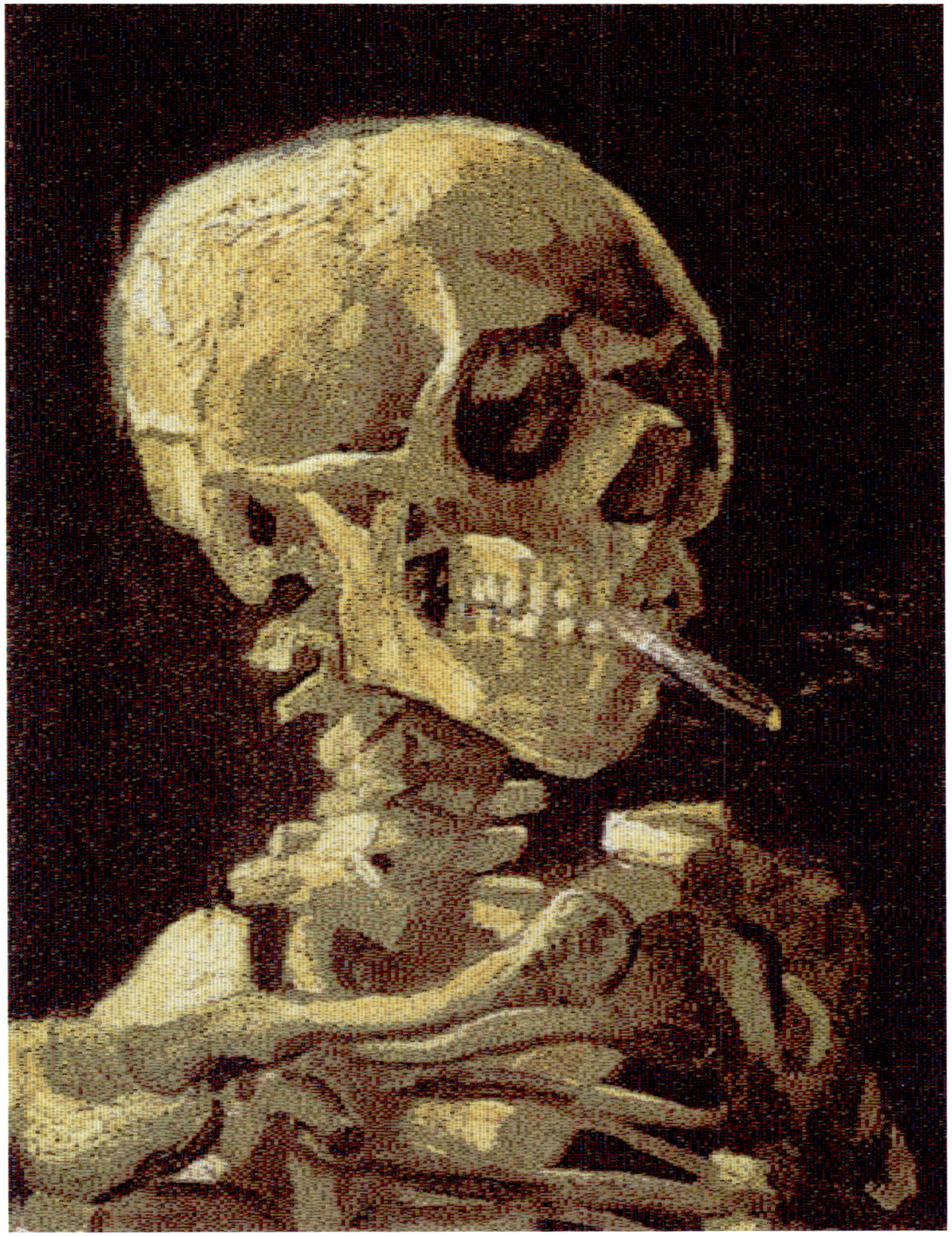

Depicts 200,000 packs of cigarettes, equal to the number of Americans who die from cigarette smoking every six months

(plate 15) Skull with Cigarette, 2008

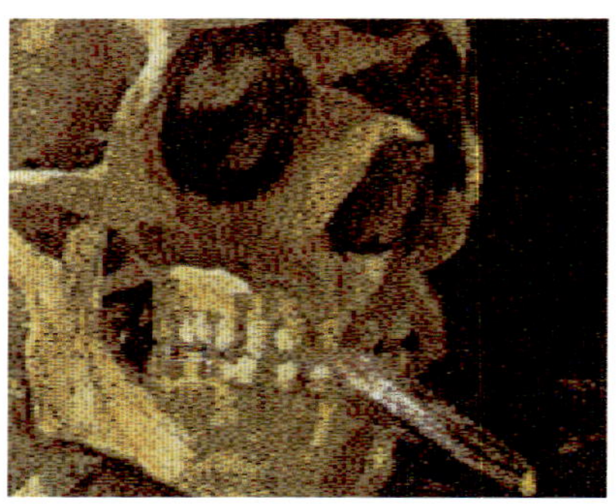
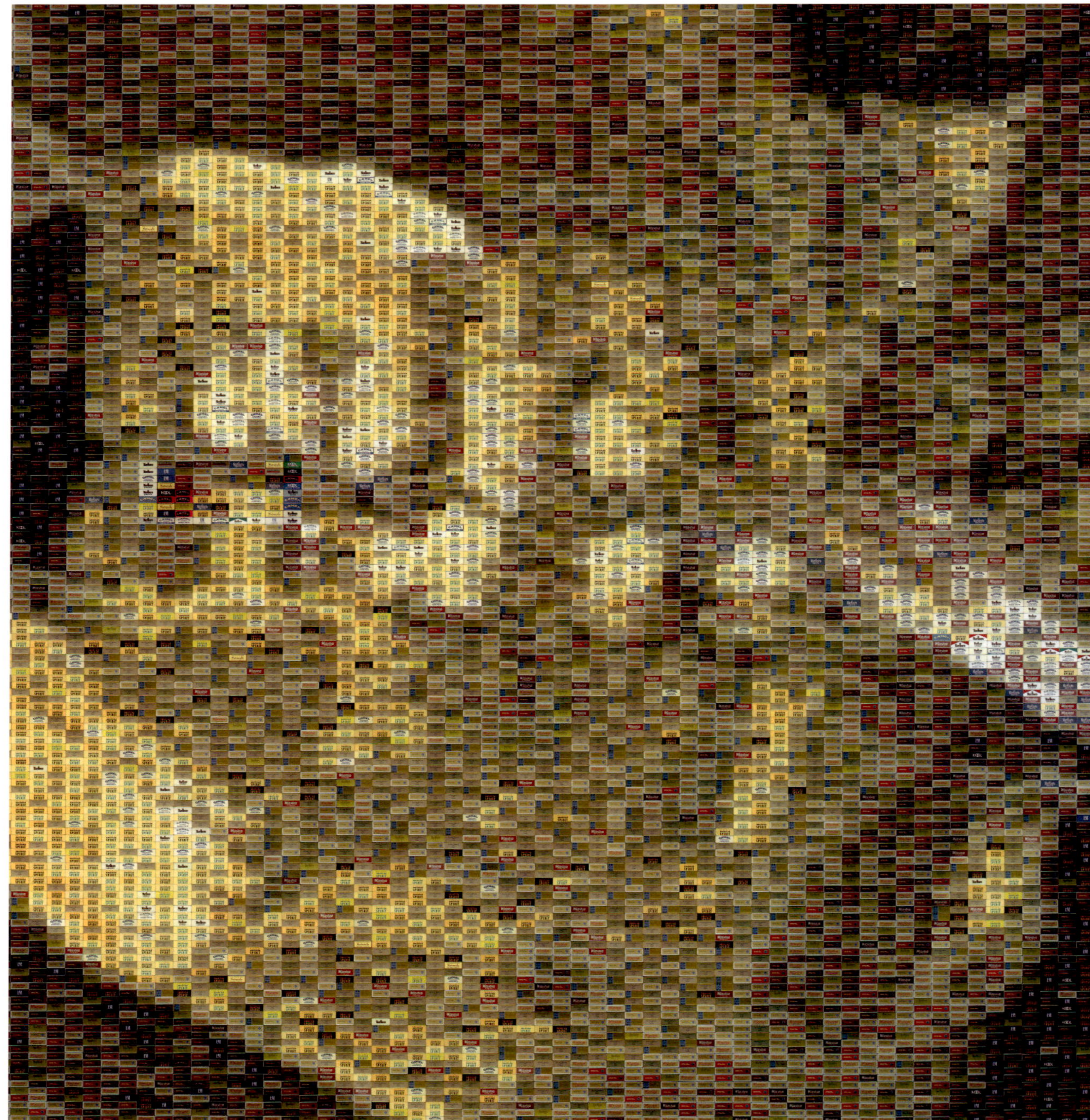

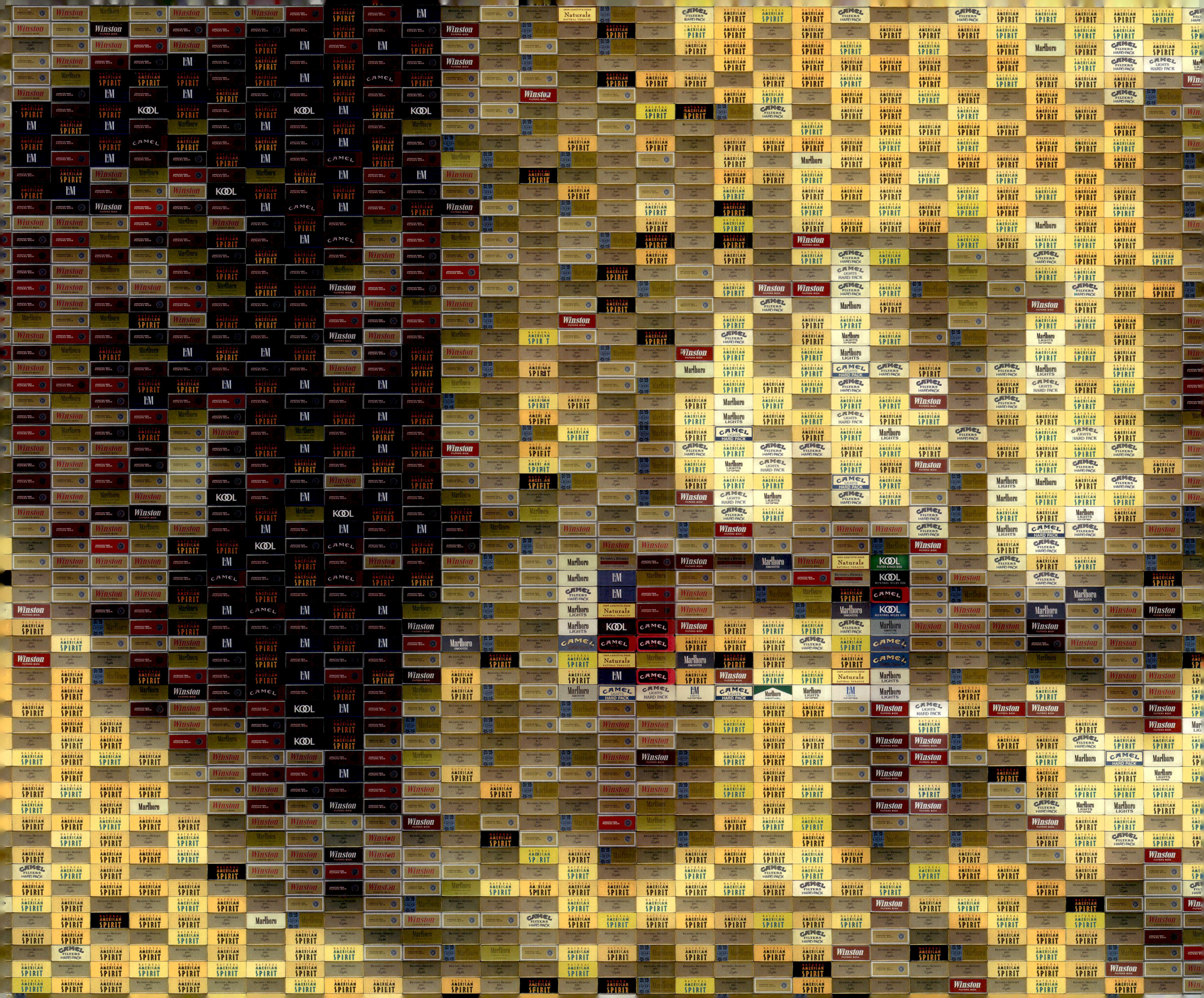

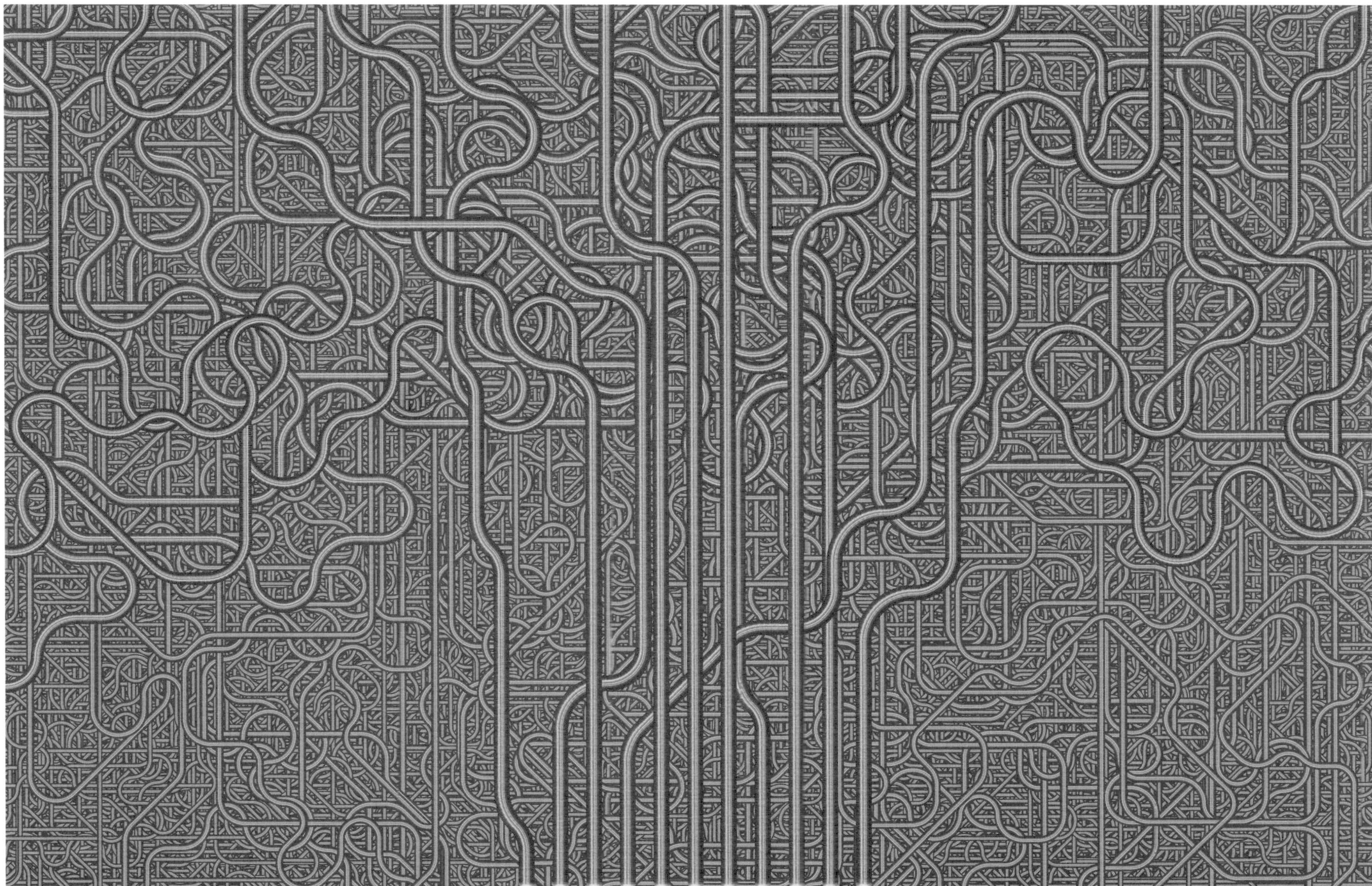

Depicts one million plastic cups, the number used on airline flights in the United States every six hours

(plate 16) Plastic Cups, 2008

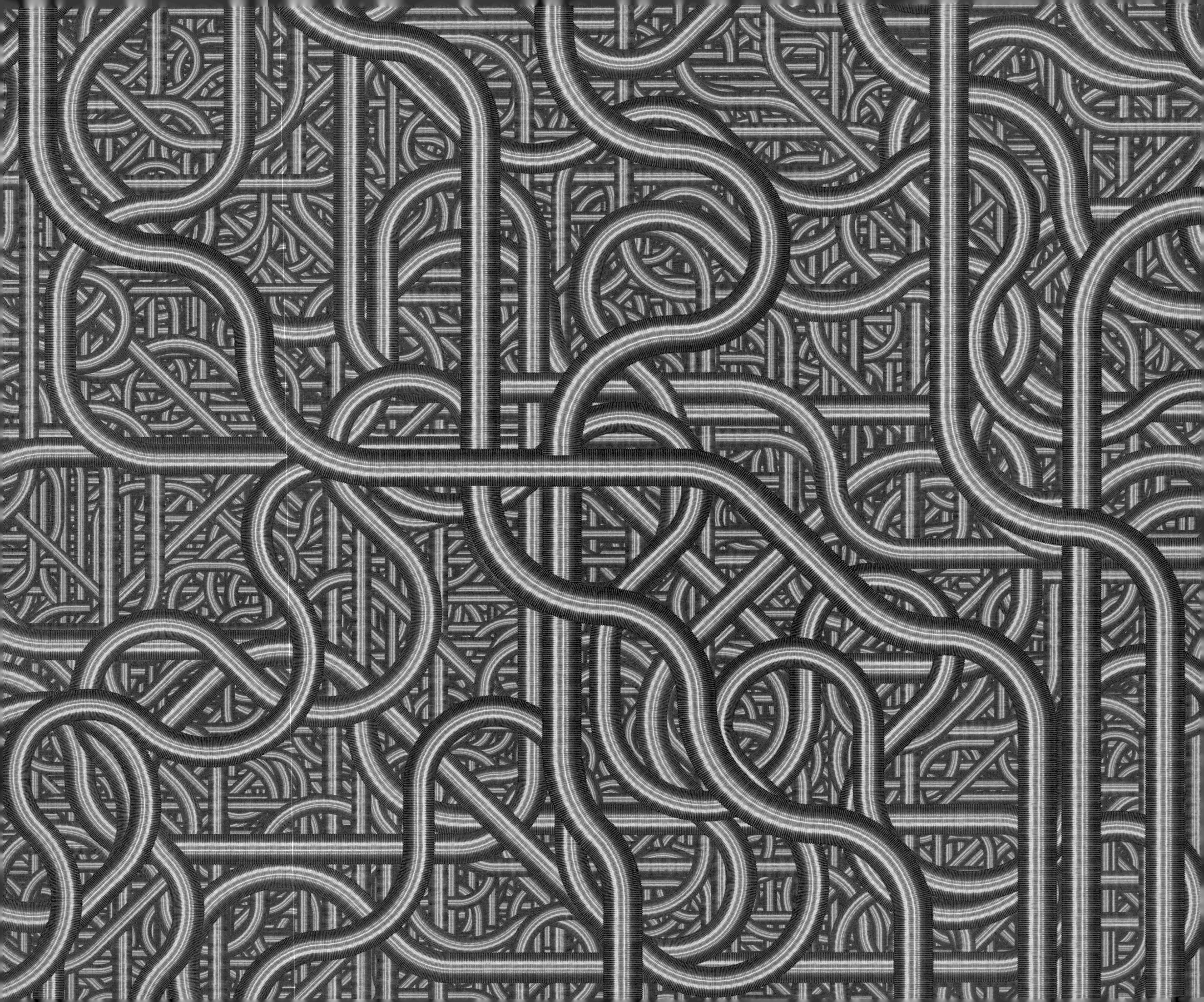

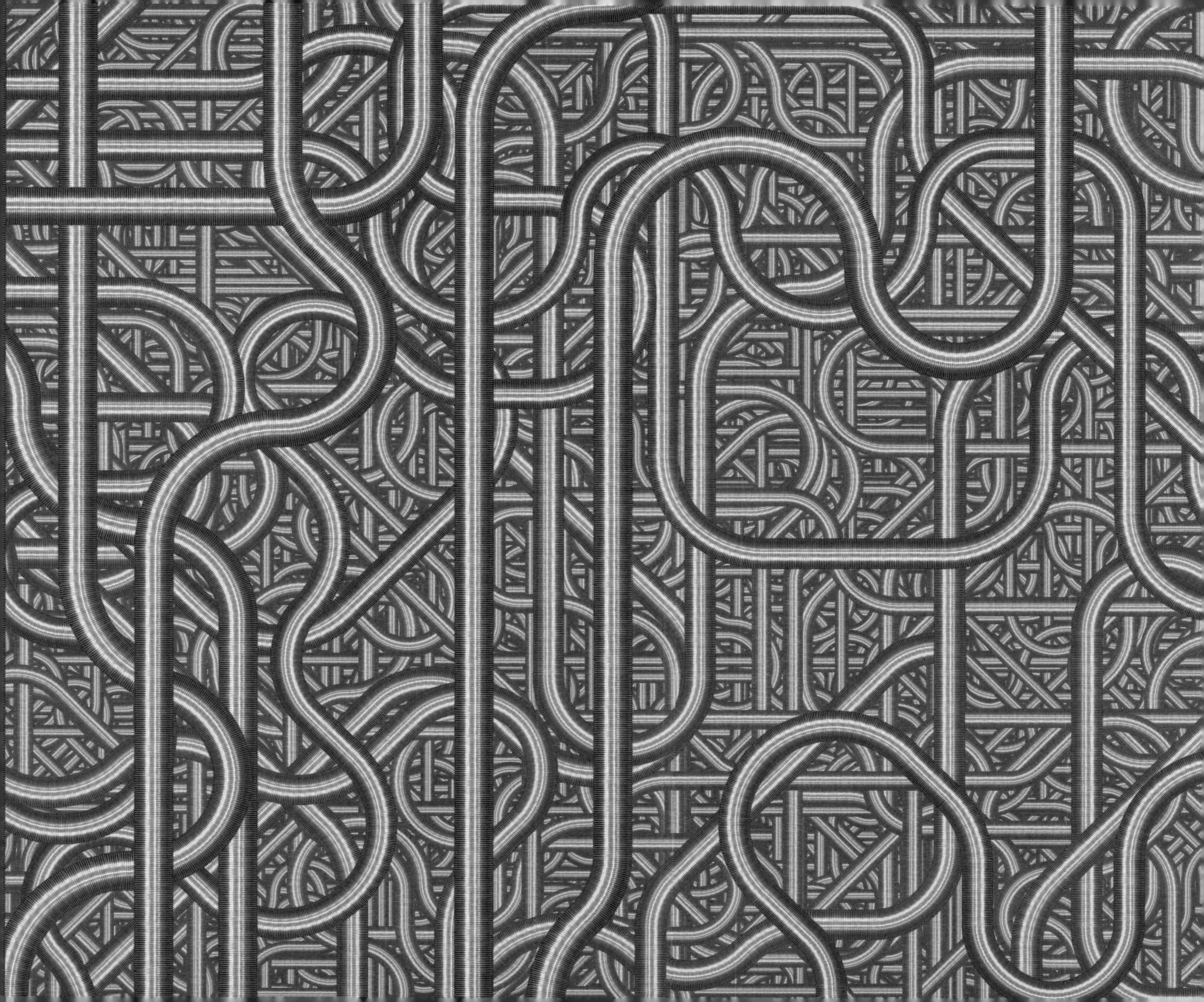

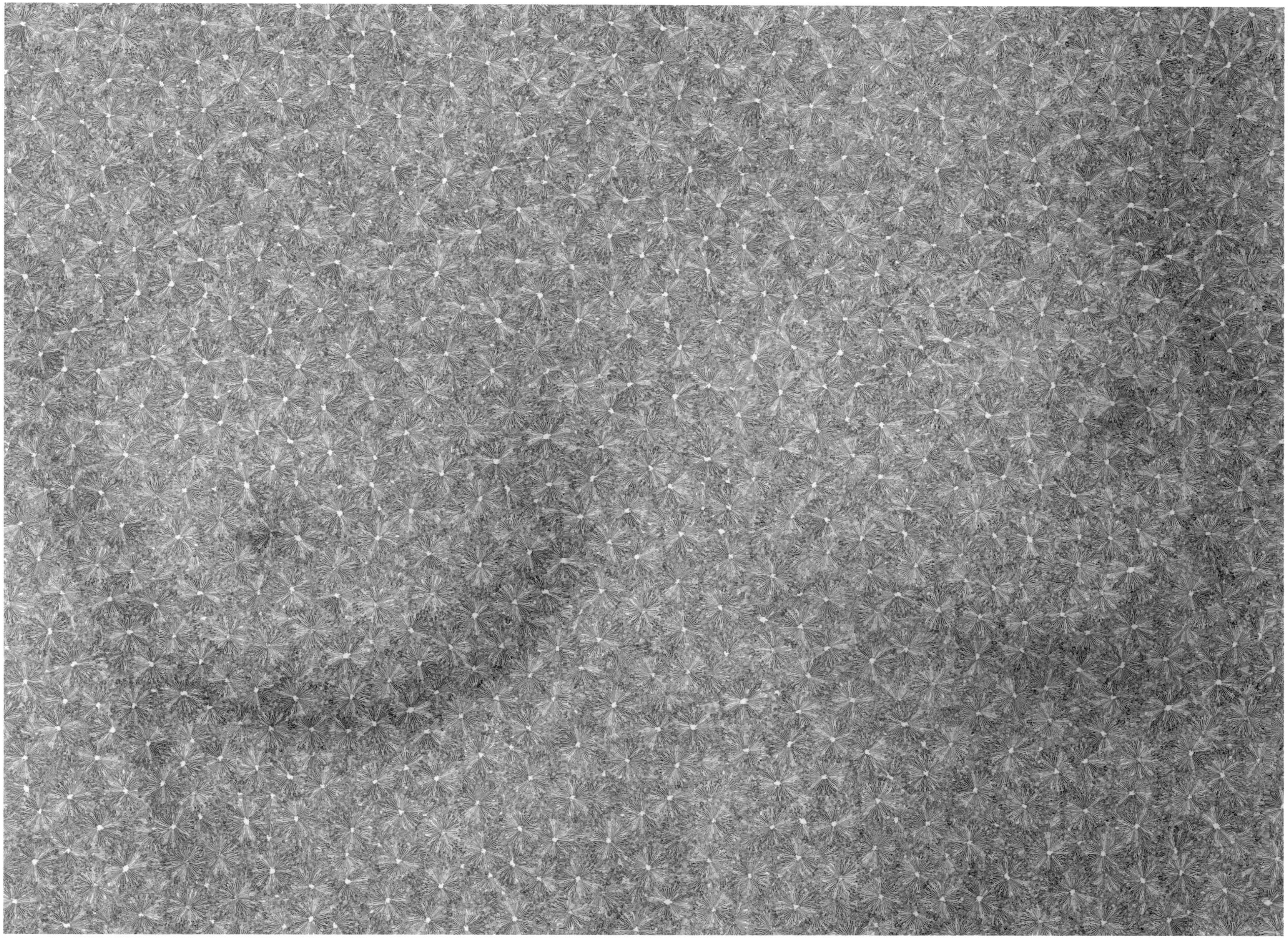

Depicts 32,000 Barbies, equal to the number of elective breast augmentation surgeries performed monthly in the United States in 2006

(plate 17) Barbie Dolls, 2008

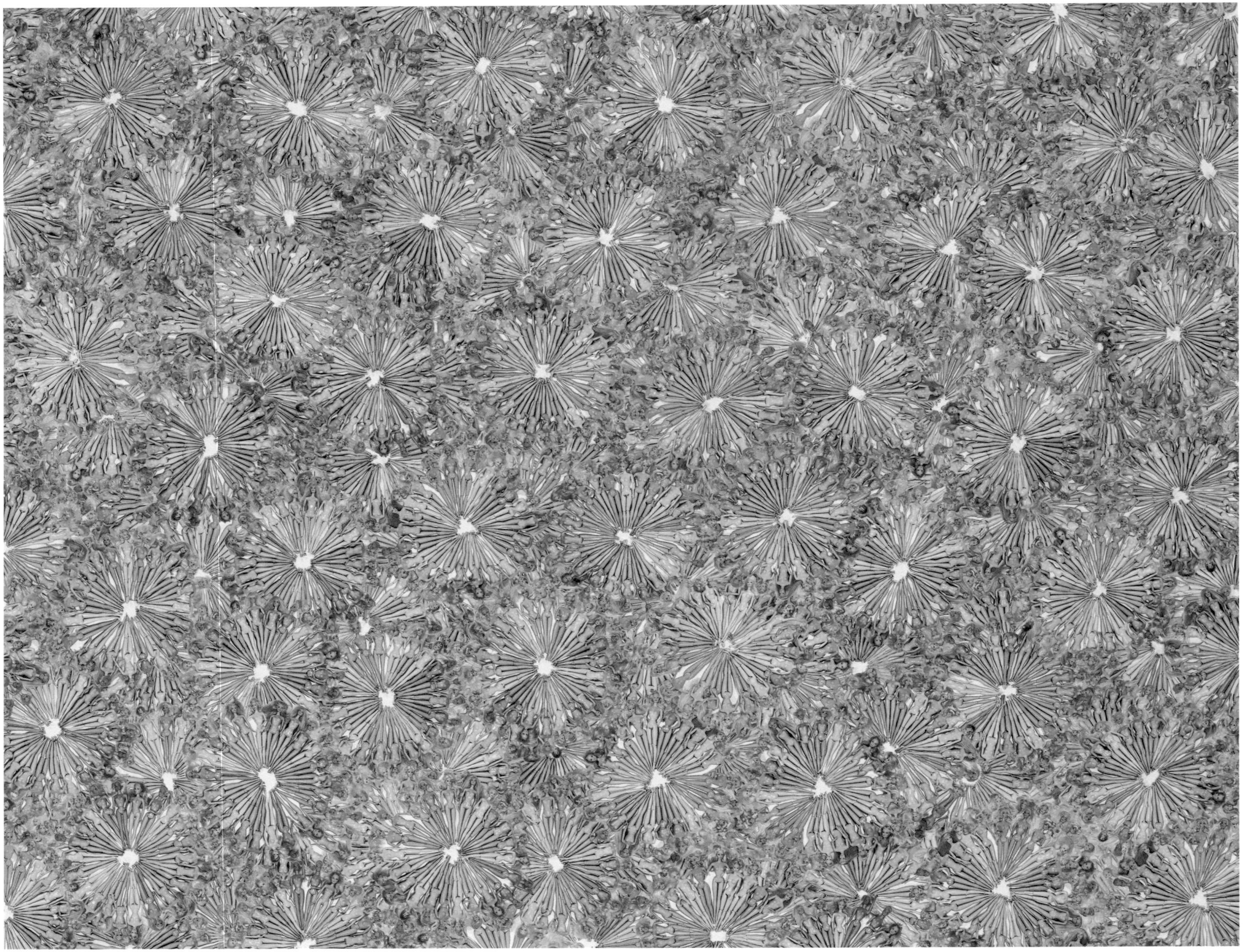

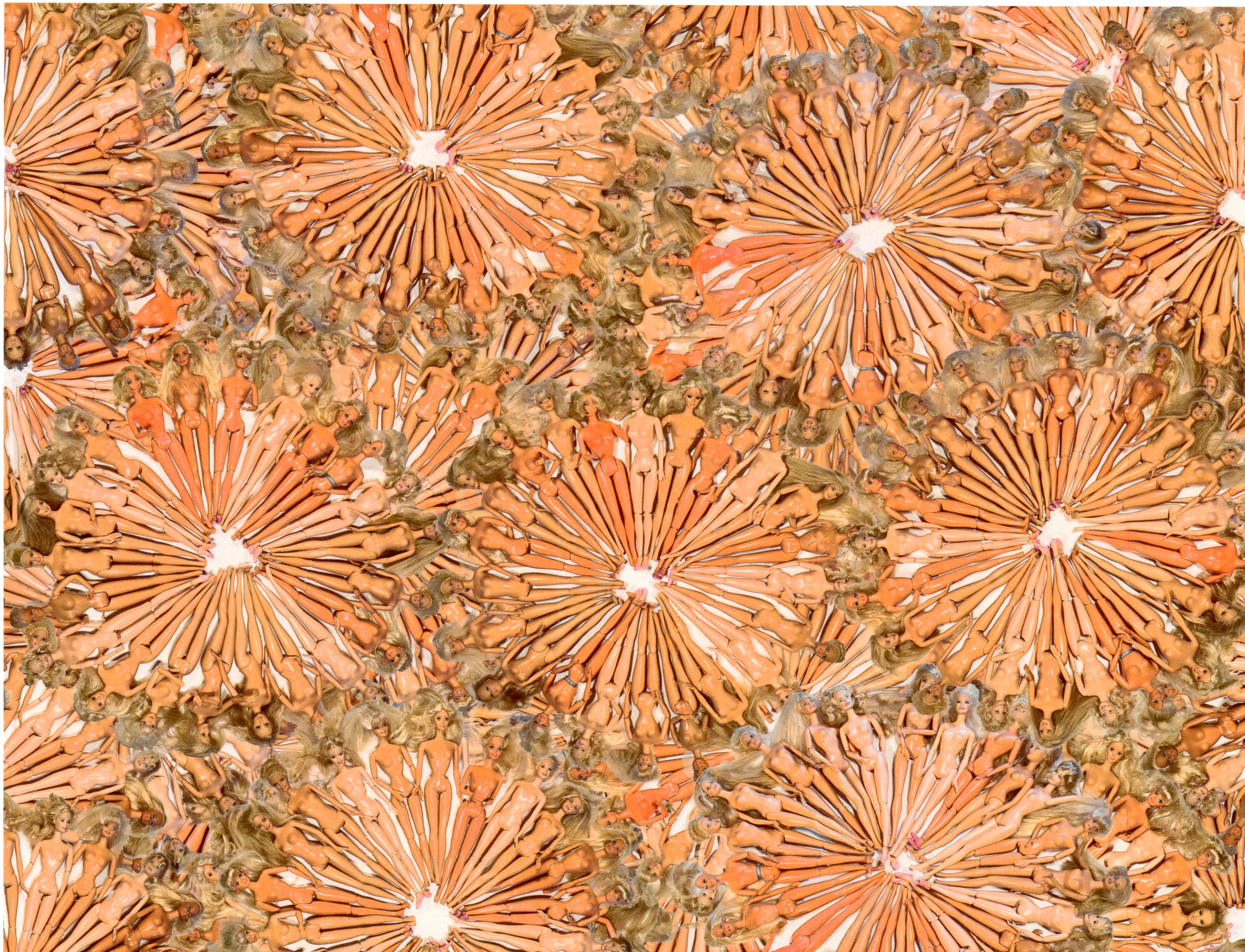

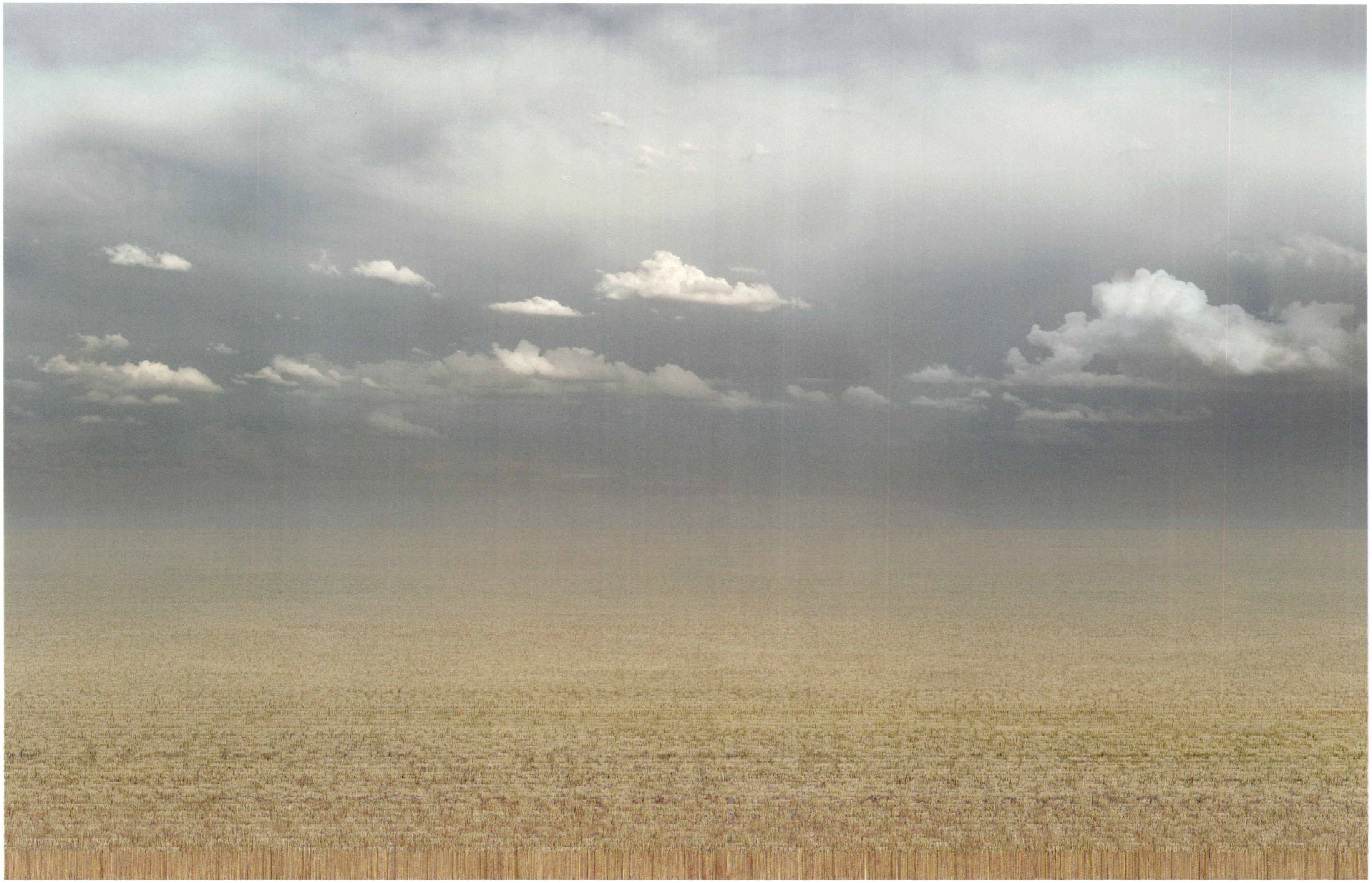

Depicts one hundred million toothpicks, equal to the number of trees cut in the United States yearly to make the paper for junk mail

(plate 18) Toothpicks, 2008

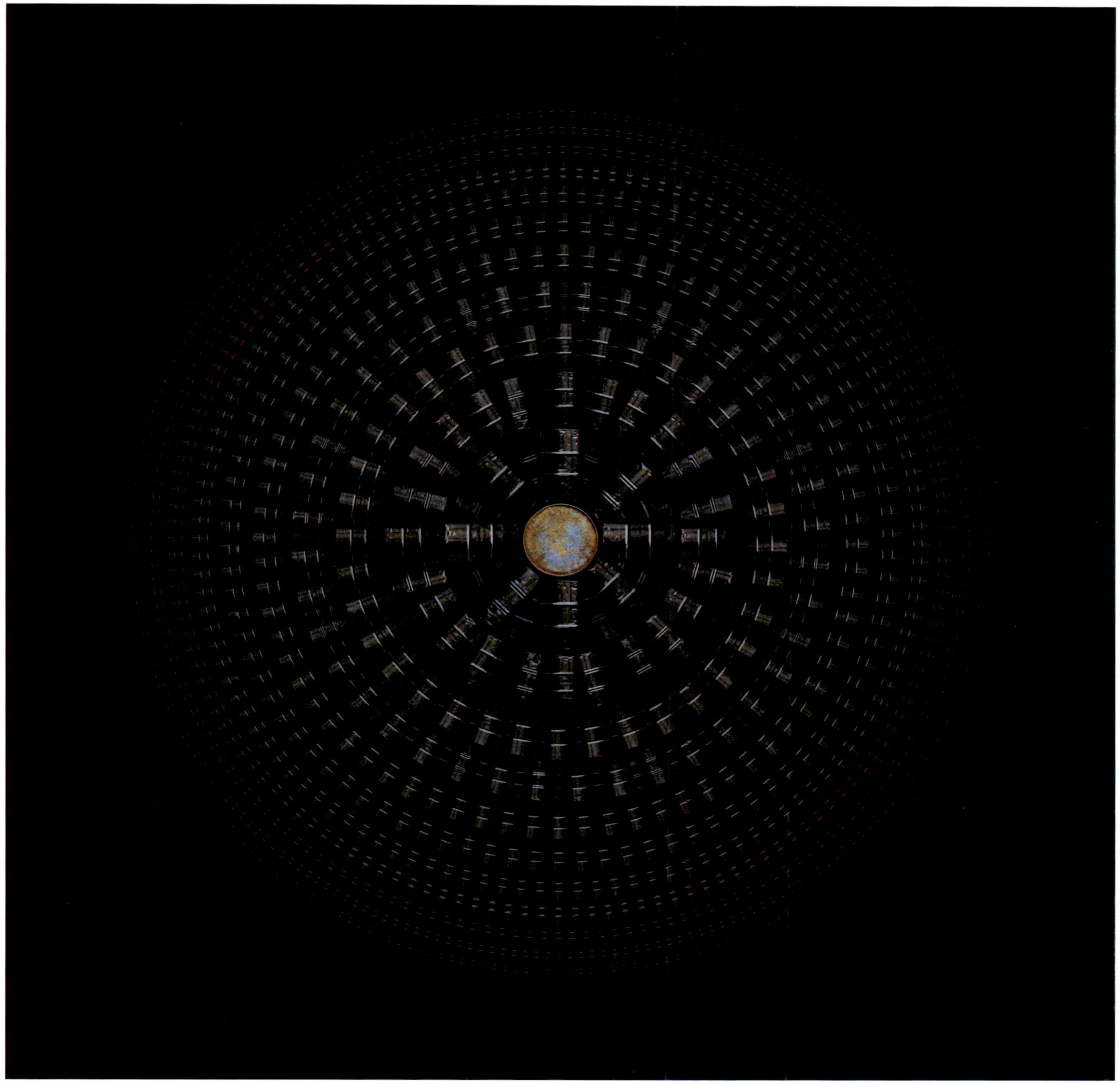

Depicts 28,000 o l barrels, equal to the volume of oil burned in the United States every two minutes

(plate 19) Oil Barrels, 20C8

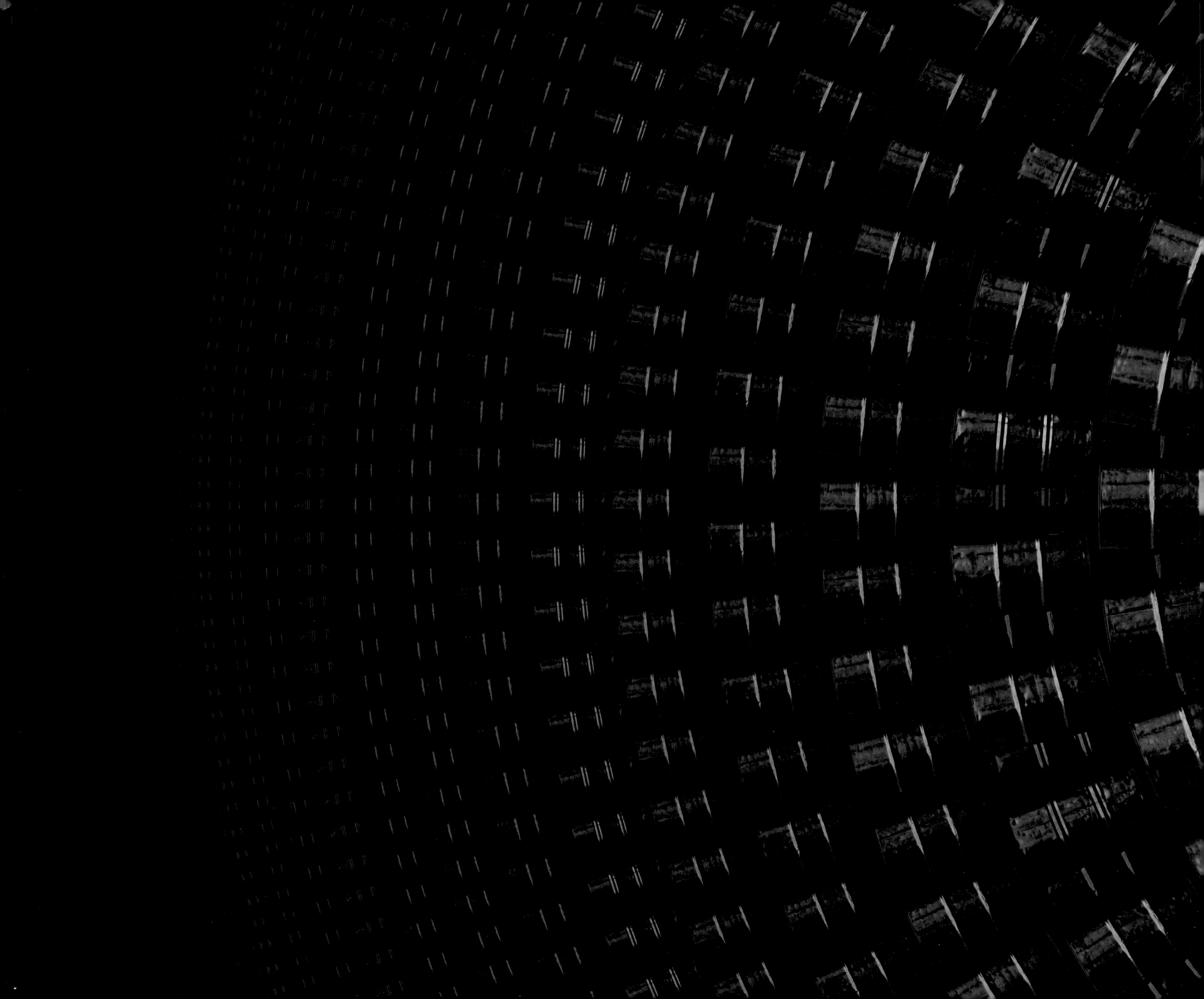

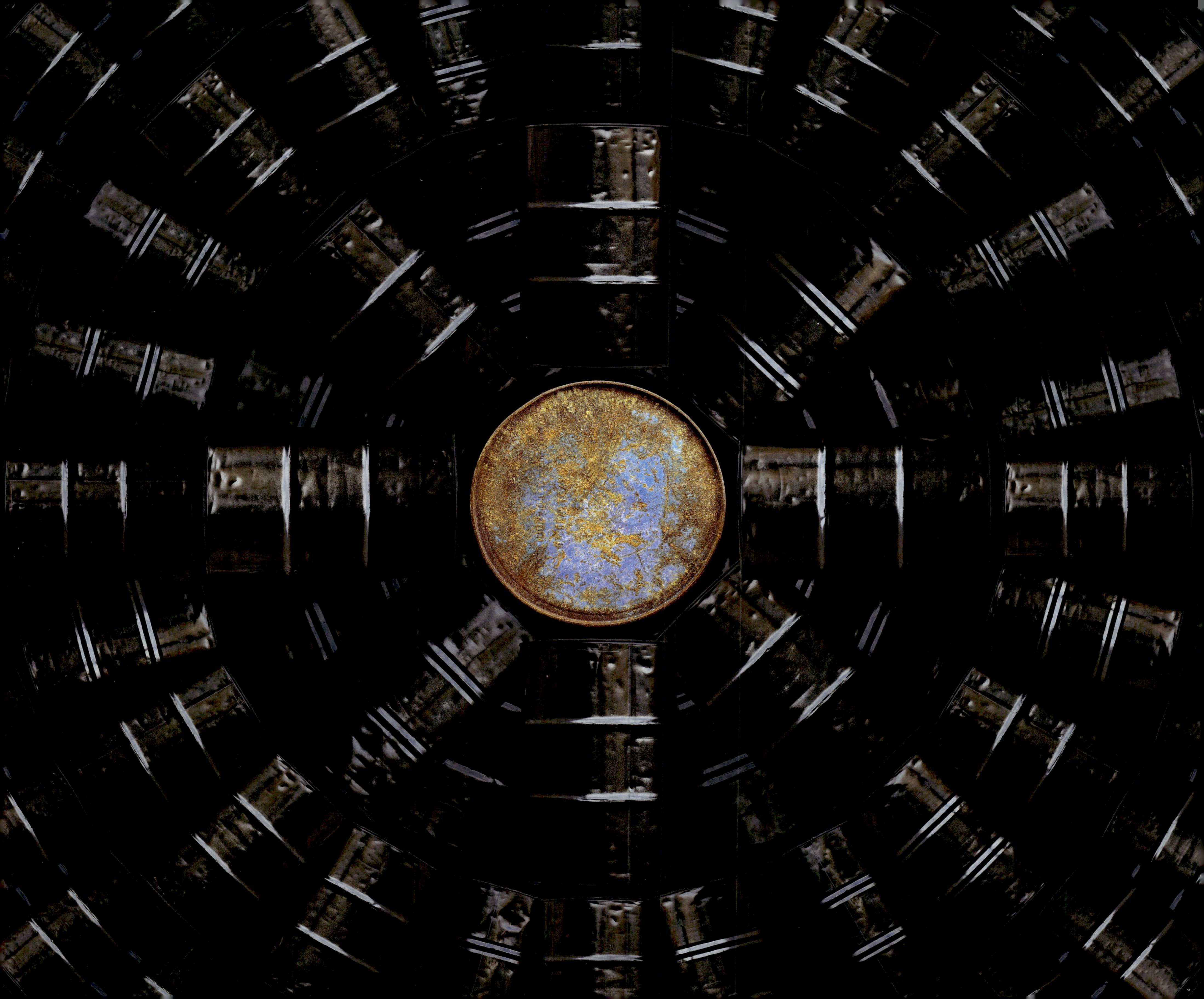

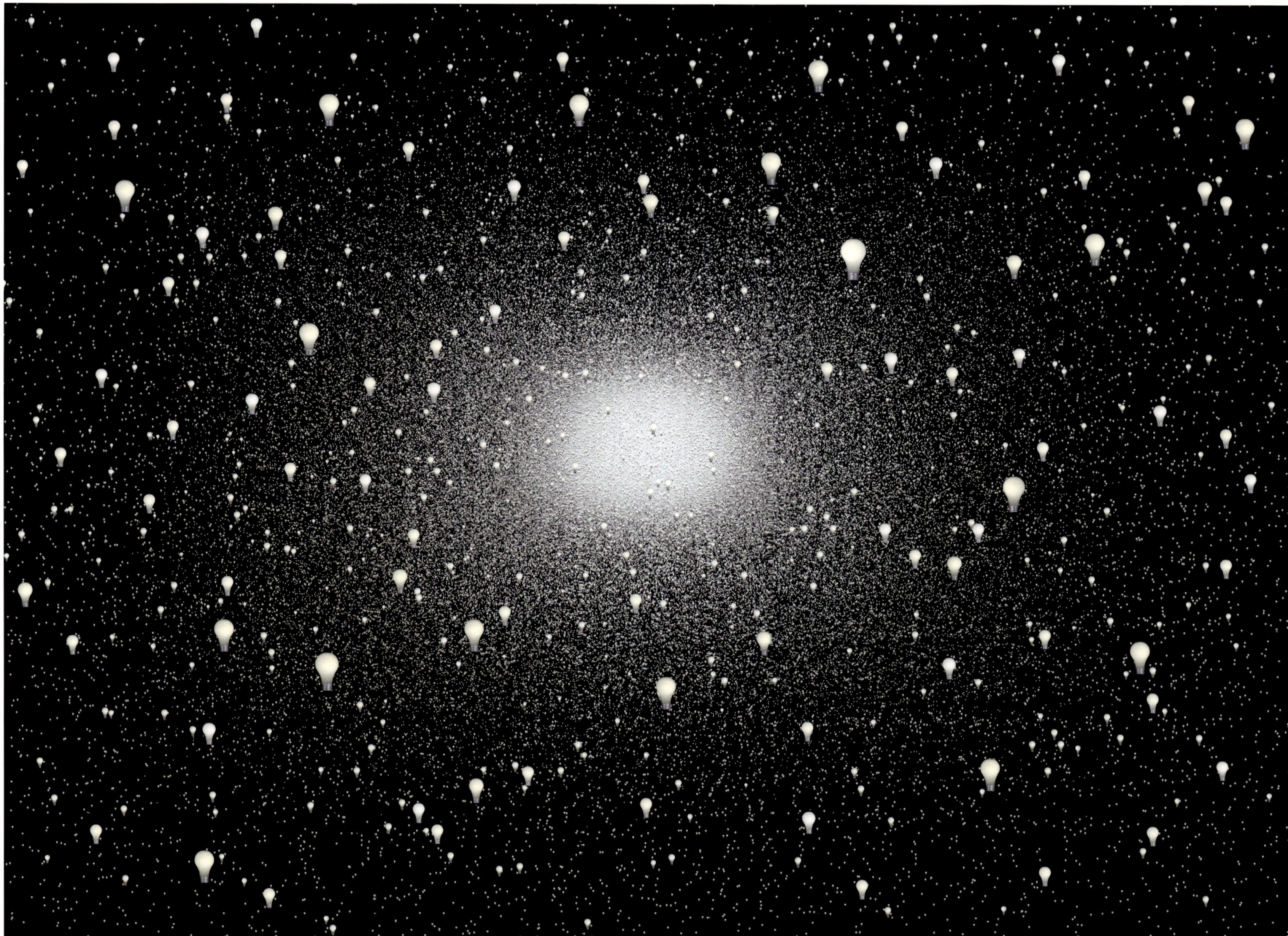

Depicts 380,000 lightbulbs, equal to the number of kilowatt hours of electricity wasted in the United States every minute by inefficient residential energy use

(plate 20) Lightbulbs, 2008

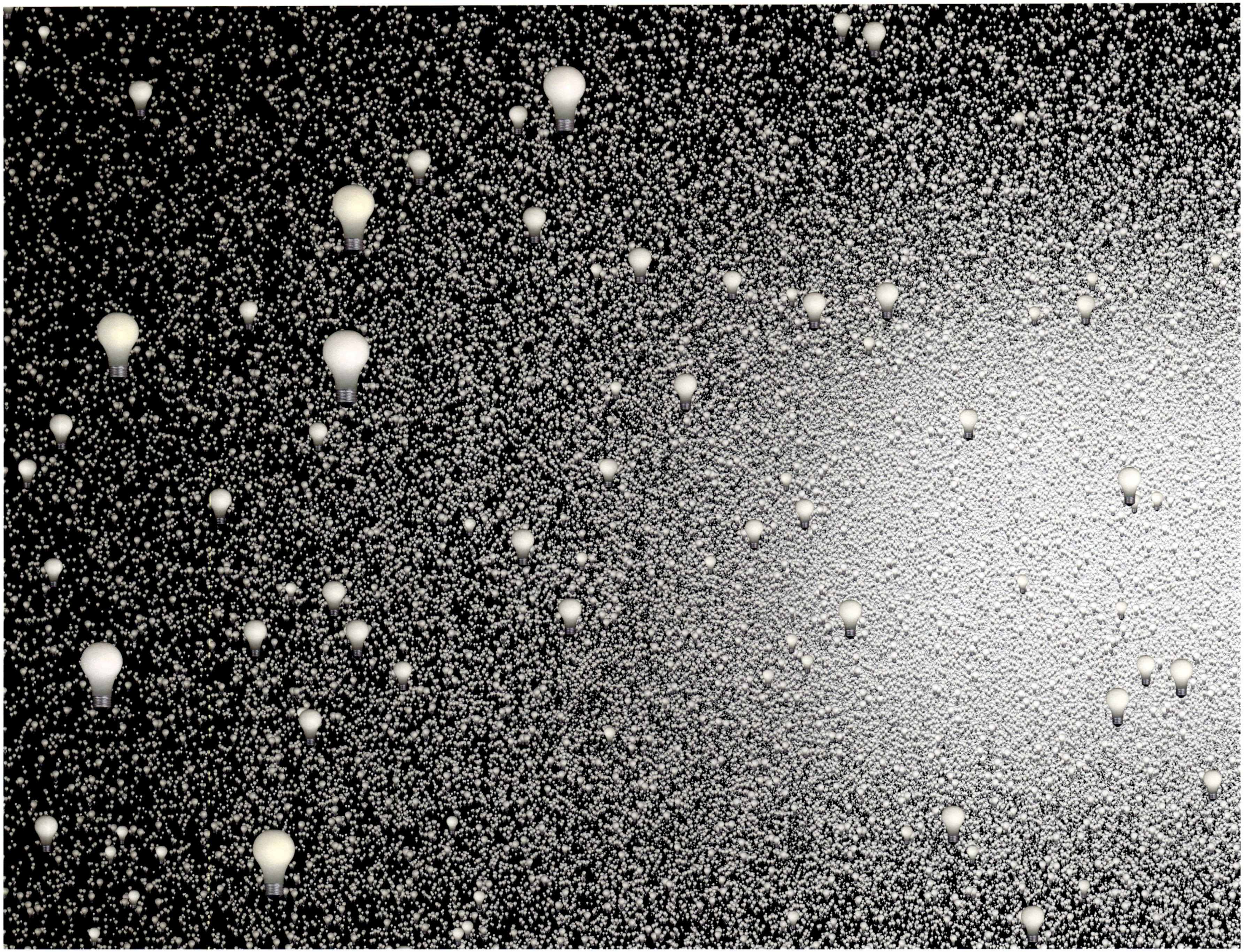

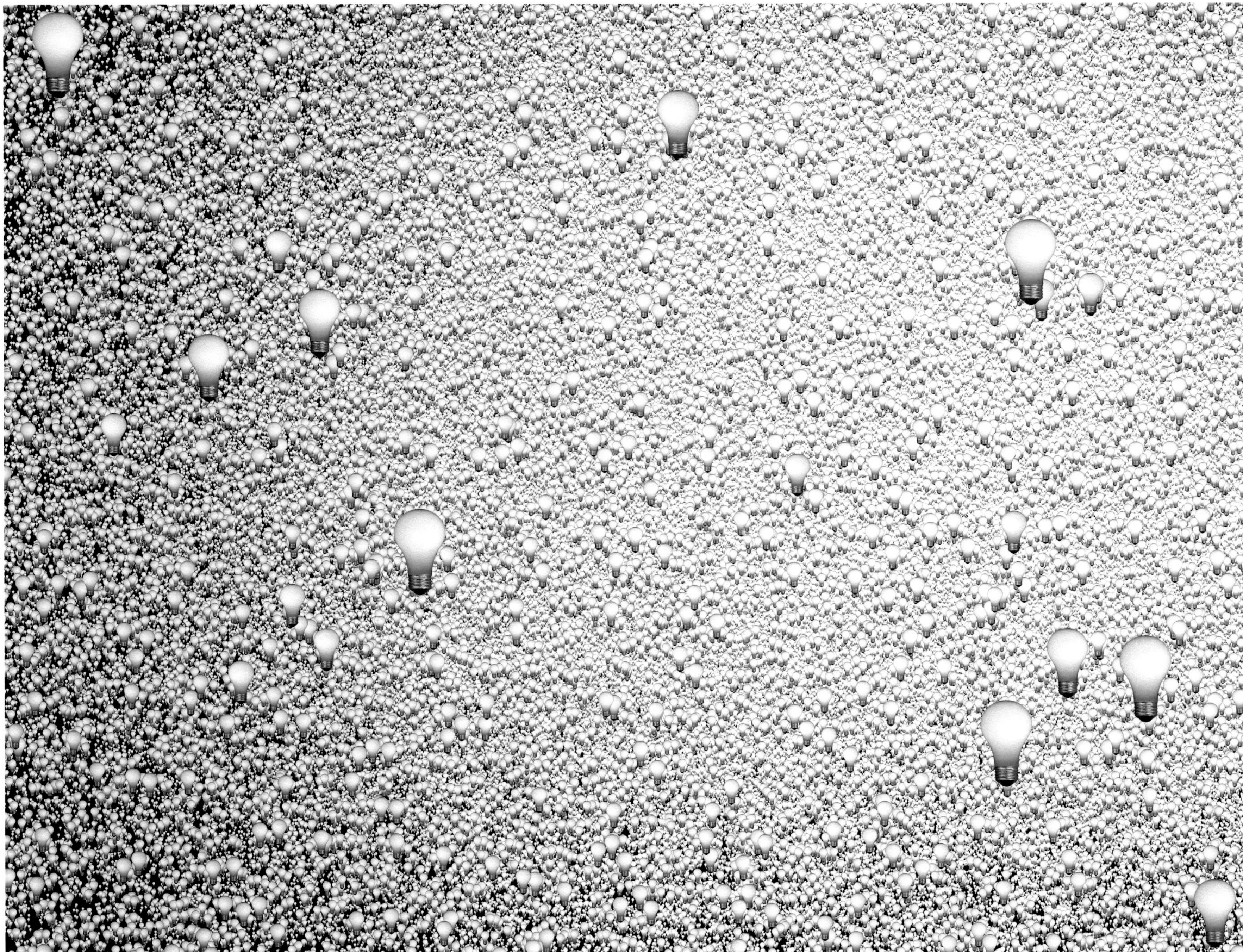

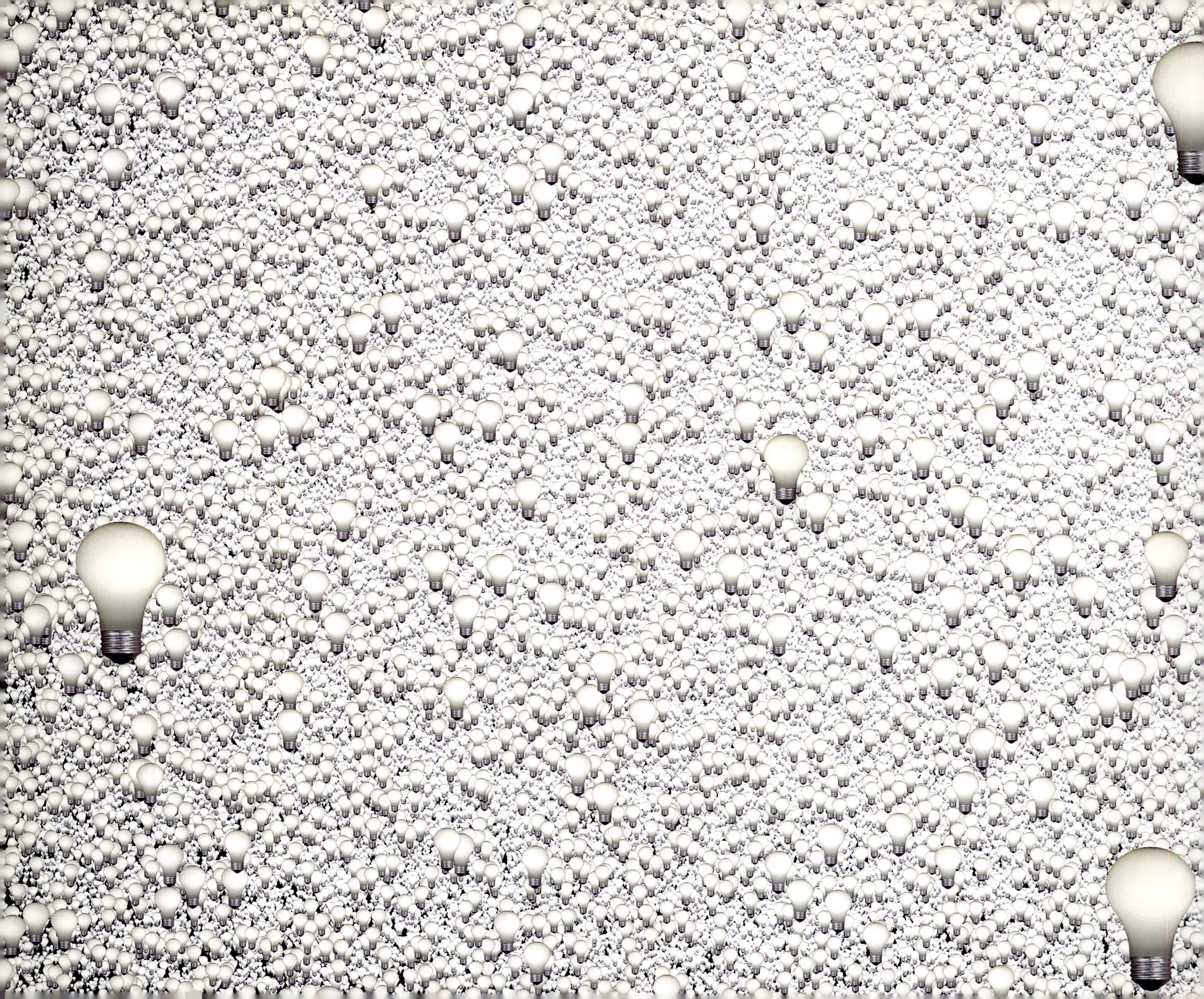

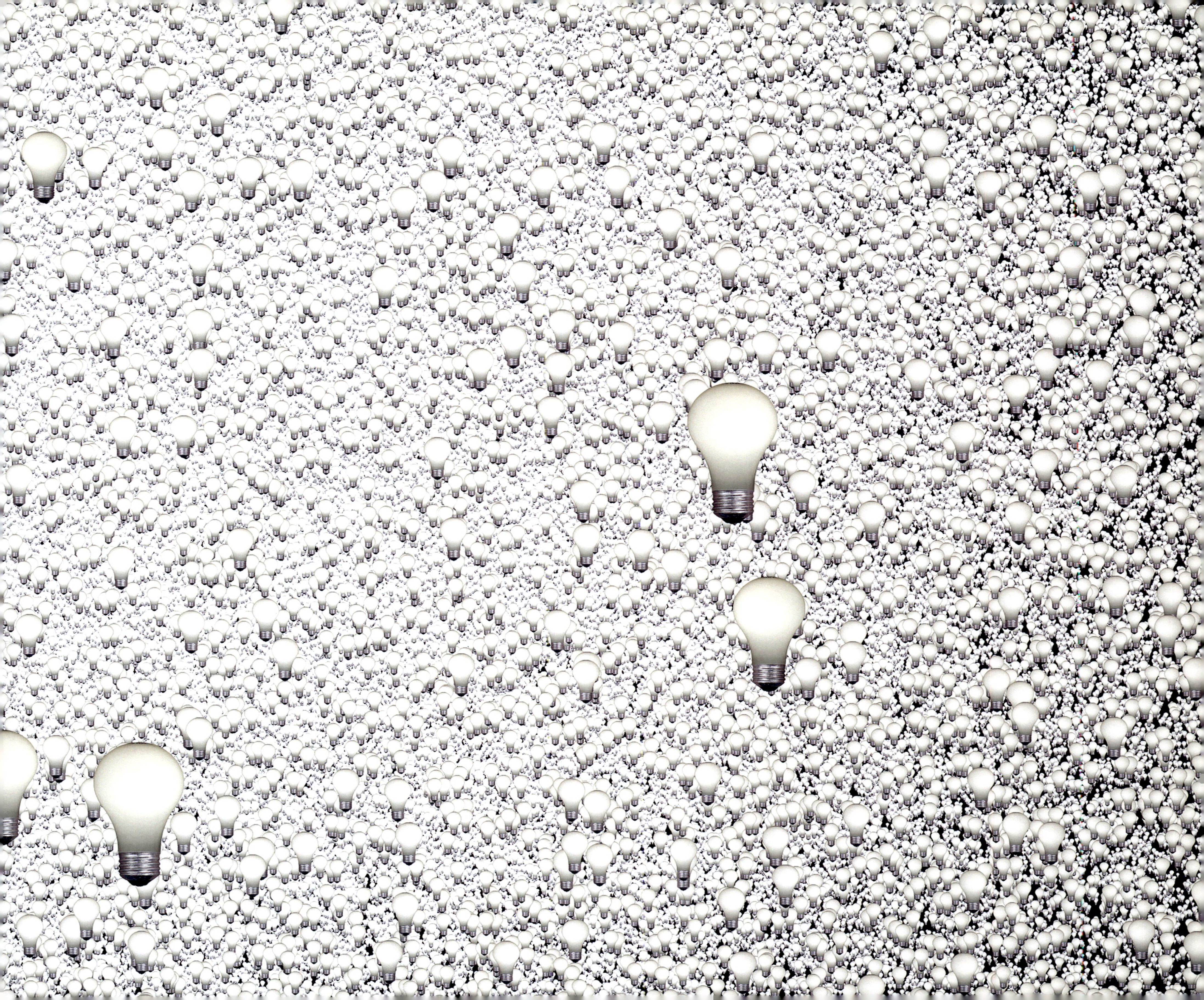

EXHIBITION CHECKLIST

INTOLERABLE BEAUTY: PORTRAITS OF AMERICAN MASS CONSUMPTION

Containers and Mt. Rainier, Tacoma, 2004
Pigmented ink-jet print
44 × 70 in.
Courtesy of the artist
(plate 2)

Mixed Recycling, Seattle, 2004
Pigmented ink-jet print
40 × 51 in.
Courtesy of the artist
(plate 3)

Sawdust, Tacoma, 2004
Pigmented ink-jet print
44 × 75 in.
Courtesy of the artist
(plate 1)

Steel Shred, Tacoma, 2004
Pigmented ink-jet print
44 × 58 in.
Courtesy of the artist
(plate 4)

RUNNING THE NUMBERS: AN AMERICAN SELF-PORTRAIT

Denali Denial, 2006
Pigmented ink-jet print
60 × 99 in.
Courtesy of the artist
Depicts 24,000 logos from the GMC Yukon Denali, equal to six weeks of sales of that model SUV in 2004
(plate 5)

Ben Franklin, 2007
Pigmented ink-jet print
126 × 102 in., in 3 vertical panels
Courtesy of the artist
Depicts 125,000 $100 bills ($12.5 million), the amount our government spent every hour on the war in Iraq from 2003 through 2008
(plate 14)

Cans Seurat, 2007
Pigmented ink-jet print
60 × 92 in.
Courtesy of the artist
Depicts 106,000 aluminum cans, the number used in the United States every thirty seconds
(plate 10)

Cell Phones, 2007
Pigmented ink-jet print
60 × 100 in.
Courtesy of the artist
Depicts 426,000 cell phones, the number retired in the United States every day
(plate 7)

Handguns, 2007
Pigmented ink-jet print
60 × 100 in.
Courtesy of the artist
Depicts 29,569 handguns, equal to the number of gun-related deaths in the United States in 2004
(plate 9)

Jet Trails, 2007
Pigmented ink-jet print
60 × 96 in.
Courtesy of the artist
Depicts 11,000 jet trails, equal to the number of commercial flights in the United States every eight hours
(plate 12)

Paper Bags, 2007
Pigmented ink-jet print
60 × 80 in.
Courtesy of the artist
Depicts 1.14 million brown paper supermarket bags, the number used in the United States every hour
(plate 11)

Plastic Bottles, 2007
Pigmented ink-jet print
60 × 120 in.
Courtesy of the artist
Depicts two million plastic beverage bottles, the number used in the United States every five minutes
(plate 6)

Prison Uniforms, 2007
Pigmented ink-jet print
120 × 276 in. (10 × 23 ft.), in 6 vertical panels
Courtesy of the artist
Depicts 2.3 million folded prison uniforms, equal to the number of Americans incarcerated in 2005
(plate 13)

Shipping Containers, 2007
Pigmented ink-jet print
60 × 100 in.
Courtesy of the artist
Depicts 38,000 shipping containers, the number processed through American ports every twelve hours
(plate 8)

Barbie Dolls, 2008
Pigmented ink-jet print
60 × 80 in.
Courtesy of the artist
Depicts 32,000 Barbies, equal to the number of elective breast augmentation surgeries performed monthly in the United States in 2006
(plate 17)

Lightbulbs, 2008
Digital C-print
72 × 96 in.
Courtesy of the artist
Depicts 380,000 lightbulbs, equal to the number of kilowatt hours of electricity wasted in the United States every minute by inefficient residential energy use
(plate 20)

Oil Barrels, 2008
Pigmented ink-jet print
64 × 64 in.
Courtesy of the artist
Depicts 28,000 oil barrels, equal to the volume of oil burned in the United States every two minutes
(plate 19)

Plastic Cups, 2008
Pigmented ink-jet print
60 × 90 in.
Courtesy of the artist
Depicts one million plastic cups, the number used on airline flights in the United States every six hours
(plate 16)

Skull with Cigarette, 2008
Digital C-print
98 × 72 in.
Courtesy of the artist
Depicts 200,000 packs of cigarettes, equal to the number of Americans who die from cigarette smoking every six months
(plate 15)

Toothpicks, 2008
Pigmented ink-jet print
60 × 99 in.
Courtesy of the artist
Depicts one hundred million toothpicks, equal to the number of trees cut in the United States yearly to make the paper for junk mail
(plate 18)

ILLUSTRATED WORKS NOT IN THE EXHIBITION

Downspouts and Cobblestones, 1992
Courtesy of the artist
(page 9)

Circuit Boards, Atlanta, 2004 (detail)
Pigmented ink-jet print
44 × 64 in.
Courtesy of the artist
(page 32)

Crushed Cars #1, Tacoma, 2004
Pigmented ink-jet print
44 × 62 in.
Courtesy of the artist
(page 20)

Drums, Seattle, 2004
Pigmented ink-jet print
40 × 50 in.
Courtesy of the artist
(page 30)

e-Bank, Tacoma 2004
Pigmented ink-jet print
44 × 59 in.
Courtesy of the artist
(page 32)

Fishing boat and remains of a home, Chalmette, 2005
from *In Katrina's Wake: Portraits of Loss from an Unnatural Disaster*
Courtesy of the artist
(page 10)

Jeep Liberty, 2005 (and detail)
Pigmented ink-jet print
120 × 88 in., in two vertical panels
Courtesy of the artist
Depicts 200,000 Jeep Liberties, the number of that model SUV sold in the United States every year
(pages 14, 15)

SELECTED EXHIBITIONS AND BOOKS BY THE ARTIST

SOLO EXHIBITIONS

2008 Winsor Gallery, Vancouver, British Columbia
Handwerker Gallery, Ithaca College, Ithaca, NY
Allen Memorial Art Museum, Oberlin College, Oberlin, OH
San Antonio Public Library and Southwest School of Art and Craft, San Antonio, TX

2007 Union Art Gallery, University of Wisconsin, Milwaukee, WI
Paul Kopeikin Gallery, Los Angeles, CA
Von Lintel Gallery, New York, NY
Armory Center for the Arts, Pasadena, CA

2006 Lannan Foundation, Santa Fe, NM
Paul Kopeikin Gallery, Los Angeles, CA
Museo Circulo de Bellas Artes, Photo España Festival, Madrid, Spain

2005 Foto&Photo Festival, Cesano Maderno, Italy
Paul Kopeikin Gallery, Los Angeles, CA
Yossi Milo Gallery, New York, NY
Photographic Center Northwest, Seattle, WA

2003 Clotworthy Arts Centre, Antrim, Ireland

SELECTED GROUP EXHIBITIONS

2008 *Digital Sequences*, Montpelier Arts Center, Laurel, MD
Waste: Artists Speak Out on Our Disposable Culture, Sharon Arts Center, Peterborough, NH
Feeling the Heat: Art, Science, and Climate Change, Deutsche Bank Gallery, New York, NY
Mountainfilm, Telluride, CO
Mass Consumption, Mesa Arts Center, Mesa, AZ
Garbage Picker, Affirmation Arts, New York, NY
3-logy Triennial 2008: Imaginative Qualities of Actual Things, Price Tower Arts Center, Bartlesville, OK

2007 *Loyal Opposition*, Texas State University Art Gallery, San Marcos, TX
The Big Picture, North Carolina Museum of Art, Raleigh, NC
Environmental Renaissance, Natural World Museum, San Francisco, CA
Envisioning Change: Melting Ice/A Hot Topic, Nobel Peace Center, Oslo, Norway; traveled to Bozar Center for Fine Arts, Brussels, Belgium; Salle d'Exposition, Monaco; and Field Museum, Chicago, IL
Green Horizons, Bates College Museum of Art, Lewiston, ME
Nature Remains: The Artist as Environmentalist, Helen Day Art Center, Stowe, VT
Weather Report, Boulder Museum of Contemporary Art, Boulder, CO
Cornucopia: Documenting the Land of Plenty, Montserrat College of Art, Beverly, MA
Altered States: Views of Transition in Recent Photography, Tufts University Art Gallery, Medord, MA
TED (Technology, Entertainment, Design) Conference, Monterey, CA

SELECTED GROUP EXHIBITIONS (Continued)

2006 International Photomedia Biennial, Society for Contemporary Photography, Kansas City, MO (Fellowship Award recipient)
Ezra and Cecile Zilkha Gallery, Wesleyan University, Middletown, CT
Lawrimore Project, Seattle, WA
Gail Gibson Gallery, Seattle, WA
Trace: Visions of Katrina, Houston Center for Photography, Houston, TX
New Orleans Museum of Art, New Orleans, LA
Port Angeles Fine Arts Center, Port Angeles, WA
Art Not Oil, London, England, traveling exhibition
Artists Responding to Violence against the Earth, Museum of Cultural Arts, Houston, TX (Best of Show award)

2005 College of Santa Fe Art Gallery, Santa Fe, NM
4th Annual Recycled Art Show, Seattle, WA
Urban Jungle Environmental Expo, U.N. International Environment Day, San Francisco, CA

2004 Allegany Arts Council Will's Creek Survey, Saville Gallery, Cumberland, MD
International Fine Art Photography Exhibition, Center for Fine Art Photography, Fort Collins, CO
Photo España Descubrimientos 04, Madrid, Spain
Artivist Film Festival, Los Angeles, CA

2000 *Intimate Landscapes*, Tacoma Art Museum, Tacoma, WA

1999 Photographic Image Gallery, Portland, OR
Earth Day, Eastshore Gallery, Bellevue, WA
Third Annual Earth Day Invitational Exhibit, Seattle, WA

1998 Second Annual Earth Day Invitational Exhibit, Seattle, WA

1997 First Annual Earth Day Invitational Exhibit, Seattle, WA
Visions of Grace, Seattle, WA

1996 Northwest Annual, Bellevue Art Museum, Bellevue, WA
The Search for Transcendence, Seattle, WA

1995 *Worth a Thousand Words*, Seattle, WA

BOOKS BY CHRIS JORDAN

In Katrina's Wake: Portraits of Loss from an Unnatural Disaster (Princeton Architectural Press, 2006)
Intolerable Beauty: Portraits of American Mass Consumption (self-published, 2005)

ACKNOWLEDGMENTS

Chris Jordan's photographic project offers a rare and unique opportunity to have a conversation about important subject matter that cuts through art-world boundaries into the environmental and social sciences. Certainly this is why he has been in such demand as a speaker, and why so many people all over the world know his images from the Web.

I thank the artist for inspiring us with his stunning and thought-provoking works of art, and for trusting us to help bring his message to new audiences through this book and the traveling exhibition.

Because of the generous funding of numerous individuals and foundations, we were able to achieve our highest goals for this project. I begin by thanking the Robert Mapplethorpe Foundation for its key early support. The Paul G. Allen Family Foundation has been a true friend of the Museum of Art, and here once again it has demonstrated its essential role in the arts of the Northwest. I am deeply obliged to Sue Coliton and Huong Vu for their advocacy of this project and to Paul Allen and Jody Patton for their unwavering encouragement. Significant funding came from the Western States Art Federation, and special thanks go to Anthony Radich. I am also grateful to the National Endowment for the Arts for its much appreciated financial assistance and thank especially Robert H. Frankel. The Compton Foundation provided important support to the artist so he could produce new work. Additional funding came from a significant contribution by Jack and Janet Creighton and from the Visual, Performing, and Literary Arts Committee at WSU. The Museum's programs are also supported by the Washington State Arts Commission and the many Friends of the Museum of Art.

I thank our museum partners for their enthusiastic responses in helping us share this important work with audiences throughout the country: at the Santa Barbara Museum of Natural History, executive director Karl Hutterer, director of education and exhibits Heather Moffatt, director of the Ty Warner Center Jenny Theodorou, and exhibits manager Simon Allen; at the Pacific Science Center in Seattle, executive director Bryce Seidl, chief financial and operating officer Darla Vivit Norris, and exhibit project coordinator Lisa Marchisio; at Haverford College, exhibitions coordinator Matthew Seamus Callinan; at the Austin Museum of Art, executive director Dana Friis-Hansen and manager of public programs Andrea Mellard; and at the Jordan Schnitzer Museum of Art at the University of Oregon, executive director Jill Hartz and associate director and curator Larry Fong.

I extend my appreciation to Jordan's studio assistant, Claire Coburn, and his Los Angeles gallerist, Paul Kopeikin.

The Museum of Art is proud of our collaboration with Marquand Books, and we appreciate the ongoing support of Ed Marquand and Adrian Lucia. Paulette Eickman provided design direction with the special support of John Hubbard, and together they created a masterful visual display. Also at Marquand, our thanks go to typesetter Maggie Lee, image manager Sara Billups, and production assistant Jeremy Linden. Our independent editor, Michelle Piranio, has been a longtime valued partner.

I am profoundly grateful to the two guest writers for this book—art critic and activist Lucy R. Lippard and environmentalist and journalist Paul Hawken. Their insights and thoughtful contributions add much to the understanding and appreciation of this remarkable work.

Finally, the staff of the Museum of Art (listed on the next page) once again demonstrated its enthusiasm and professionalism, taking on the challenge of securing funding for the book and exhibition, handling the loans, coordinating the transportation, packing, and presentation of the artworks as well as seeing to all the details of the traveling exhibition. Truly, without each and every member of the staff working together for the highest ideals of art, there would be no Chris Jordan exhibition, no book, and no program of distinction at Washington State University. They deserve the highest praise for another job well done.

My sincere thanks to all.

Chris Bruce
Director, Museum of Art / Washington State University

MUSEUM OF ART STAFF

Chris Bruce, *Director*

Jill Aesoph, *Director of Development*
Brenna Helm-Manwaring, *Administrative Assistant*
Zach Mazur, *Assistant to the Curator*
Heather McGeachy, *Installation Technician/Security*
Tonya Murray, *Program Coordinator*
Dustin Price, *Installation Technician*
Camille Rigby, *Media and Public Relations Coordinator*
Anna-Maria Shannon, *Associate Director*
Debby Stinson, *Graphic Designer*
Keith Wells, *Curator of Exhibitions and Collections*
Seth Echlin, Adam Johnston, Marlene Kittilson, Kiry Oeurn, Kayla Sheeley, and Julia White, *Gallery Attendants*

MUSEUM OF ART DOCENTS

Marlene Guse
Sally Horton
Lolly Owens
Marion Park
Alma Rocha
Sharon Wells
H. Clare Wiser

This publication was produced in conjunction with the exhibition *Chris Jordan: Running the Numbers,* organized by Chris Bruce for the Museum of Art/Washington State University, Pullman.

Museum of Art/Washington State University, Pullman, Washington
January 14–April 4, 2009

Santa Barbara Museum of Natural History, Santa Barbara, California
May 15–September 11, 2009

Pacific Science Center, Seattle, Washington
October 3, 2009–January 3, 2010

Haverford College, Haverford, Pennsylvania
January 22–March 5, 2010

Austin Museum of Art, Austin, Texas
May 22–August 15, 2010

Jordan Schnitzer Museum of Art, University of Oregon, Eugene, Oregon
January–March 2011

Distributed by
Prestel Verlag
Königinstrasse 9, 80539 Munich, Germany
Tel. (89) 24 29 08-300; Fax (89) 24 29 08-335
www.prestel.de

900 Broadway, Suite 603, New York, NY 10003, USA
Tel. (212) 995-2720; Fax (212) 995-2733

This exhibition and publication are made possible by the generous support of:
The Paul G. Allen Family Foundation
Compton Foundation
Jack and Janet Creighton
The Robert Mapplethorpe Foundation
National Endowment for the Arts
Visual, Performing, and Literary Arts Committee, WSU
Washington State Arts Commission
Western States Art Federation

The Museum's exhibition program is supported by the Friends of the Museum of Art.

Library of Congress Cataloging-in-Publication Data
Jordan, Chris, 1963–
Running the numbers: an American self-portrait/Chris Jordan; introduction by Chris Bruce; essays by Lucy R. Lippard and Paul Hawken.
p. cm.
"Produced in conjunction with the exhibition "Chris Jordan: Running the Numbers," organized by Chris Bruce for the Museum of Art/Washington State University, Pullman"—T.p. verso.
ISBN 978-0-9755662-5-1
1. United States—Environmental conditions—Pictorial works—Exhibitions. 2. United States—Social conditions—Pictorial works—Exhibitions. 3. Consumption (Economics)—Social aspects—United States—Pictorial works—Exhibitions. 4. Packaging waste—United States—Pictorial works—Exhibitions. 5. Refuse and refuse disposal—United States—Pictorial works—Exhibitions. 6. United States—Statistics—Pictorial works—Exhibitions. 7. Jordan, Chris, 1963-—Exhibitions. 8. Photography, Artistic—Exhibitions. I. Washington State University. Museum of Art. II. Title.
GE150.J67 2009
779'.936300973—dc22 2008050846

Designed by Paulette Eickman and John Hubbard
Edited by Michelle Piranio
Proofread by Laura Iwasaki and Leslie Eliel
Typeset by Maggie Lee
Color management by iocolor, Seattle
Produced by Marquand Books, Inc., Seattle
www.marquand.com
Printed and bound in China by Artron Color Printing, Ltd.

Cover: Plastic Bottles, 2007 (detail of plate 6)